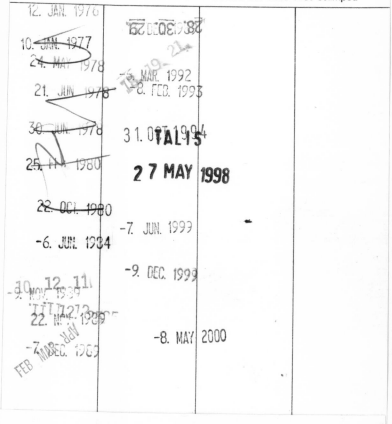

FLUID FLOW

for Chemical Engineers

FLUID FLOW
for Chemical Engineers

F. A. HOLLAND

Professor of Chemical Engineering, University of Salford
and
Partner in Salchem Associates, Consulting Chemical Engineers

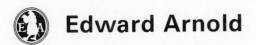 Edward Arnold

First published 1973
by Edward Arnold (Publishers) Ltd.
25 Hill Street
London W1X 8LL

ISBN: 0 7131 3300 7 Boards ✓
ISBN: 0 7131 3301 5 Paper

Printed in Great Britain by J. W. Arrowsmith Ltd, Bristol

Preface

This book is a basic undergraduate text in fluid flow. It is a summary of the fluid flow content of the chemical engineering degree course at the University of Salford. The book is written throughout in SI units and is divided into two parts. Part 1 is a conventional treatment of fluid flow and contains a minimum of mathematics. Part 1 is suitable for use in Higher National Certificate and Higher National Diploma courses in chemical engineering. Part 2 makes use of vector analysis and more sophisticated mathematics. Part 2 deals with the flow of Newtonian liquids with reference to rectangular and cylindrical coordinate systems. The treatment of non-Newtonian flow in rectangular and cylindrical coordinate systems requires the use of tensors. Tensors are only used in Master's degree courses at Salford and are consequently omitted from this text. It can readily be seen that the transport phenomena approach used in Part 2 is far more powerful than the largely empirical approach used in Part 1. Nevertheless a clear understanding of physical boundary conditions and the engineering aspects of a problem are essential if the transport phenomena approach is to be used effectively. Parts 1 and 2 together are suitable for use in honours degree courses in chemical engineering. Part 1 and much of Part 2 are also suitable for use in ordinary degree courses in chemical engineering.

The material in this book is also used in the one week refresher courses in fluid mechanics, which are periodically run by the Department of Chemical Engineering at the University of Salford, for the Institution of Chemical Engineers. It is hoped that the book will also be useful for chemists, mechanical engineers and other technical people concerned with the flow of fluids.

The author believes that there is no substitute for wide reading in a subject. However, this can be done more effectively with reference to a basic framework. This book, which is largely a collection of lecture notes with the emphasis on brevity, is designed to provide such a framework.

The author would like to express his gratitude to Miss Barbara Buckley for typing the manuscript and to his colleague, Mr F. A. Watson, for checking the material and reworking the calculations. He also greatly appreciates the valuable help given by Mr J. Swolkein and Mr P. Diggory with the drawings.

F A HOLLAND

Contents

Chapter 3   Flow of incompressible non-Newtonian fluids in pipes

Chapter 4   Pumping of liquids

Chapter 5   Mixing of liquids in tanks

Chapter 6   Flow of compressible fluids in conduits

Chapter 7   Flow of two phase gas liquid mixtures in pipes

Chapter 8   Flow measurement

Chapter 9   Fluid motion in the presence of solid particles

Chapter 10   Introduction to unsteady state fluid flow

Part Two   Vector Methods in Fluid Flow

Chapter 11   Vector methods in fluid flow and the equations
of continuity and momentum transfer

Chapter 12   Applications of modified Navier Stokes equations in rectangular coordinates

Chapter 13   Applications of modified Navier Stokes equations in horizontal cylindrical coordinates

Chapter 14   Applications of modified Navier Stokes equations in vertical cylindrical coordinates

List of example calculations

xiii

Nomenclature

a	blade width, m
A	area, m^2
b	width, m
C	Chezy coefficient $\sqrt{g/j_f}$, $m^{\frac{1}{2}}/s$
C	constant, usually dimensionless
C	solute concentration, kg/m^3
C_d	discharge or drag coefficient, dimensionless
C_p	heat capacity per unit mass at constant pressure, J/(kg K)
C_v	heat capacity per unit mass at constant volume, J/(kg K)
d	diameter, m
d_{ep}	equivalent diameter of annulus $D_i - d$ for pressure drop, m
d_{eq}	equivalent diameter of annulus $(D_i^2 - d_0^2)/d_0$ for heat transfer, m
D	diameter, m
$\dfrac{D}{Dt}$	substantial time derivative, $\dfrac{\partial}{\partial t} + v_x \dfrac{\partial}{\partial x} + v_y \dfrac{\partial}{\partial y} + v_z \dfrac{\partial}{\partial z}$ in Cartesian co-ordinates, s^{-1}
E	efficiency function $\left(\dfrac{1}{P_A/V}\right)\left(\dfrac{1}{t}\right)$, m^3/J
E	total energy per unit mass, J/kg or m^2/s^2
f	Fanning friction factor, dimensionless
F	energy per unit mass required to overcome friction, J/kg
F	force, N
g	gravitational acceleration, 9.81 m/s^2
G	mass flow rate, kg/(s m^2)
h	head, m
H	height, m
H	enthalpy per unit mass, J/kg
$\mathbf{i}$	unit vector, dimensionless
I_T	tank turnovers per unit time in equation (5.2–6), s^{-1}
$\mathbf{j}$	unit vector, dimensionless
j_f	friction factor, dimensionless
k	exponent in equation (6.2–8), dimensionless
k	proportionality constant in equation (5.1–1), dimensionless
$\mathbf{k}$	unit vector, dimensionless

K	consistency coefficient, $kg/(s^{2-n}m)$
K	parameter in equation (2.5–3), dimensionless
K	proportionality constant in equation (2.9–10), dimensionless
K_c	parameter in Carmen Kozeny equation, dimensionless
K_p	consistency coefficient for pipe flow, $kg/(s^{n-2}m)$
KE	kinetic energy flow rate, W
L	length of pipe or tube, m
L	mixing length in equation (2.9–8), m
ln	$\log_e$, dimensionless
log	$\log_{10}$, dimensionless
m	mass of fluid, kg
m	number, dimensionless
M	flow rate of fluid, kg/s
(MW)	molecular weight, kg/kmol
n	number, dimensionless
n	power law index, dimensionless
n'	flow behaviour index in equation (3.1–1), dimensionless
N	rotational speed, rev/s
N_C	compressibility factor in equation (6.2–5), dimensionless
N_{FR}	Froude number, dimensionless
N_{HE}	Hedstrom number, dimensionless
N_M	Mach number, dimensionless
N_P	power number, dimensionless
N_{RE}	Reynolds number, dimensionless
N_{WE}	Weber number, dimensionless
N_Y	yield number for Bingham plastics, dimensionless
$NPSH$	net positive suction head, m
p	pitch, m
P	pressure, N/m^2
P_A	agitator power, W
P_B	brake power, W
P_E	power, W
q	heat energy per unit mass, J/kg or m^2/s^2
Q	volumetric flow rate, m^3/s
r	blade length, m
r	pressure ratio, dimensionless
r	radius, m
r_f	recovery factor in equation (6.6–8), dimensionless
R	shear stress, N/m^2
R_G	gas constant, 8.3143 kJ/(kmol K)
s	distance, m
s	scale reading in equation (8.5–1), dimensionless
s	slope $\sin\theta$, dimensionless
s	parameter in Laplace transform, s^{-1}
S	cross-sectional flow area, m^2
S_o	surface area per unit volume, m^{-1}
t	time, s
T	temperature, K
T_o	stagnation temperature in equation (6.6–7), K
TS	tip speed, m/s
u	mean linear velocity, m/s
u_p	terminal settling or falling velocity, m/s
U	internal energy per unit mass, J/kg or m^2/s^2
v	linear velocity, m/s

V	volume, m^3
V	volume per unit mass, m^3/kg
w	weight fraction, dimensionless
W	work energy per unit mass, J/kg or m^2/s^2
x	distance, m
x	exponent in equation (5.4–5), dimensionless
x_v	volume concentration of solids, dimensionless
X_{tt}	Lockhart Martinelli parameter in equation (7.2–8), dimensionless
y	distance, m
y	exponent in equation (5.4–5), dimensionless
Y_1	expansion factor in equation (6.7–16), dimensionless
z	distance, m
α	velocity distribution factor in equation (1.6–7), dimensionless
α	reciprocal of holding time Q/V, s^{-1}
γ	ratio of heat capacities C_p/C_v, dimensionless
$\dot{\gamma}$	shear rate, s^{-1}
δ	thickness of boundary layer, m
ε	roughness of pipe, m
ε	voidage fraction, dimensionless
η	kinematic viscosity, m^2/s
η	efficiency factor in equation (5.2–2), dimensionless
θ	angle or slope, dimensionless
μ	dynamic viscosity of fluid, $kg/(s\ m)$ or $N\ s/m^2$
ρ	density of fluid, kg/m^3
σ	surface tension, N/m
τ	torque, N m
ϕ	power function in equation (5.4–6), dimensionless
ϕ	Lockhart Martinelli parameter in equation (7.2–9), dimensionless
ψ	correction factor in equation (9.1–10), dimensionless
ω	angular velocity, rad/s
Δe	per cent error in equation (8.5–5), dimensionless

$$\nabla \quad \text{del, } \mathbf{i}\frac{\partial}{\partial x} + \mathbf{j}\frac{\partial}{\partial y} + \mathbf{k}\frac{\partial}{\partial z} \text{ in Cartesian coordinates, } m^{-1}$$

$$\nabla^2 \quad \text{Laplacian operator, } \frac{\partial^2}{\partial x^2} + \frac{\partial^2}{\partial y^2} + \frac{\partial^2}{\partial z^2} \text{ in Cartesian coordinates, } m^{-2}$$

Subscripts

a	referring to apparent
A	referring to agitator
b	referring to packed bed
B	referring to yield stress
c	referring to coarse suspension, coil, contraction, or critical
d	referring to discharge side
D	referring to displacement
e	referring to eddy, equivalent, or expansion
f	referring to friction
G	referring to gas
i	referring to inside of pipe or tube
L	referring to liquid
m	referring to manometer liquid, mean, or a number

M	referring to mixing
n	referring to a number
o	referring to outside of pipe or tube or a reference level
p	referring to pipe or solid particle
r	referring to reduced
s	referring to sonic, stream, suction side, or system
t	referring to time or transient
T	referring to tank or total
vp	referring to vapour
V	referring to volume
w	referring to pipe or tube wall
W	referring to water

Part one BASIC FLUID FLOW

1
Fluids in motion

1.1 Units and dimensions

Mass, length and time are commonly used primary units. Their dimensions are written as M, L and T respectively. Other units are derived in terms of mass, length and time. In the Système International d'Unités, commonly known as the SI system of units, the primary units are the kilogram kg, the metre m and the second s. A number of derived units are listed in Table (1.1–1).

Although the SI unit for the amount of substance is the mole, the kmol has been used in this text for convenience and consistency.

TABLE (1.1–1)

quantity	derived unit	symbol	relationship to primary units
force	newton	N	kg m/s²
work, energy, quantity of heat	joule	J	N m
power	watt	W	J/s
area	square metre		m²
volume	cubic metre		m³
density	kilogram per cubic metre		kg/m³
velocity	metre per second		m/s
acceleration	metre per second squared		m/s²
pressure	newton per square metre		N/m²
surface tension	newton per metre		N/m
dynamic viscosity	newton second per metre squared		N s/m² or kg/(s m)

1.2   Flow patterns

In general, fluids in motion have different velocities at different points in a line perpendicular to the direction of flow. The particular distribution of velocities depends on the nature of the flow which in turn is a function of the geometry of the container, the physical properties of the fluid, and its mass flow rate.

For the most part, flow can be characterized either as laminar or as turbulent flow.

Laminar flow.   This is also called viscous or streamline flow. In this type of flow, layers of fluid move relative to each other without any macroscopic intermixing. Laminar flow systems are commonly represented graphically by streamlines. There is no fluid flow across these lines. A velocity distribution results from shear stresses which in turn are present because of viscous frictional forces.

Turbulent flow.   In turbulent flow, there is an irregular random movement of fluid in directions transverse to the main flow. This irregular fluctuating motion can be regarded as superimposed on the mean motion.

Consider fluid flow with reference to an ordinary rectangular Cartesian coordinate system x, y, z. A point velocity at any instant in the x direction can be written as

$$v_x = \bar{v}_x + \bar{v}'_x$$

where $\bar{v}_x$, the mean point velocity, is defined as

$$\bar{v}_x = \frac{1}{\Delta t} \int_0^{\Delta t} v_x \, dt \qquad (1.2\text{--}1)$$

In equation (1.2–1), Δt is a time interval which need be only a few seconds, since the irregular fluctuations are extremely rapid. If the mean velocity $\bar{v}_x$ is constant with time, the motion in the x direction is said to be in steady state. If motions exist in the y and z directions, they can similarly be expressed as the sum of a mean and a fluctuating velocity.

1.3   Newton's law of viscosity and momentum transfer

Consider two parallel plates of area A distance dz apart shown in Figure (1.3–1). The space in between the plates is filled with a fluid. The lower plate travels with a velocity v and the upper plate with a velocity $v - dv$. The small difference in velocity dv between the plates results in a resisting force F acting over the plate area A due to viscous frictional effects in the fluid.

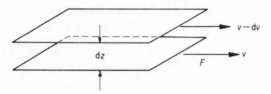

Figure (1.3–1)
Shear between two plates.

Thus a force F must be applied to the lower plate to maintain the difference in velocity dv between the two plates.

The force per unit area F/A is known as the shear stress R.

Since the velocity v decreases as the distance z increases, the velocity gradient is written with a negative sign as $-dv/dz$.

Newton's law of viscosity states that the shear stress R is proportional to the velocity gradient $-dv/dz$ in the fluid. The constant of proportionality is known as the coefficient of dynamic viscosity μ. Newton's law of viscosity can be written

$$R = -\mu \frac{dv}{dz} \qquad (1.3–1)$$

Fluids which obey this equation are called Newtonian fluids. Fluids which do not obey this equation are called non-Newtonian fluids.

In terms of velocity in the horizontal x direction, equation (1.3–1) can be rewritten for a point in the z direction in the form

$$R_{zx} = -\mu \frac{dv_x}{dz} \qquad (1.3–2)$$

or for a point in the radial r direction in the form

$$R_{rx} = -\mu \frac{dv_x}{dr} \qquad (1.3–3)$$

Newton's law of viscosity holds for Newtonian fluids in streamline flow. For Newtonian fluids in streamline flow, the velocity gradient $-dv/dz$ is also the shear rate conventionally written as $\dot{\gamma}$.

Newton's law of viscosity is commonly written in one of the following three forms:

$$R = \mu\dot{\gamma}$$ (1.3–4)

$$\dot{\gamma} = \frac{R}{\mu}$$ (1.3–5)

or

$$\mu = \frac{R}{\dot{\gamma}}$$ (1.3–6)

i.e.

$$\text{dynamic viscosity} = \frac{\text{shear stress}}{\text{shear rate}}$$

In a fluid in laminar flow, fast moving molecules diffuse into
slow moving streams and vice versa, resulting in a transfer of
momentum in a direction perpendicular to the direction of flow.
The rate of momentum transfer is the same as the shear stress R_{zx}
given by equation (1.3–2).

Equation (1.3–2) may also be written as

$$R_{zx} = -\eta\rho\frac{\mathrm{d}v_x}{\mathrm{d}z}$$ (1.3–7)

where $\eta = \mu/\rho$, the viscous diffusivity or kinematic viscosity.

In turbulent flow, momentum transfer takes place by the movement of eddies imposed on the ordinary molecular motion. The
rate of momentum transfer through regions of turbulent flow is
given by the equation

$$R_{zx} = -(\eta + \eta_e)\rho\frac{\mathrm{d}v_x}{\mathrm{d}z}$$ (1.3–8)

where η_e is the eddy viscous diffusivity. In turbulent flow, the eddy
viscous diffusivity η_e is much greater than the molecular viscous
diffusivity η. Thus large shear stresses exist in turbulent fluids.

1.4   Non-Newtonian behaviour

For Newtonian fluids a plot of shear stress R against shear rate $\dot{\gamma}$
on Cartesian coordinates is a straight line having a slope equal to
the coefficient of dynamic viscosity μ. For many fluids a plot of

R against $\dot{\gamma}$ does not give a straight line. These are the so-called non-Newtonian fluids. Plots of R against $\dot{\gamma}$ are experimentally determined using a viscometer.

The term viscosity has no meaning for a non-Newtonian fluid unless it is related to a particular shear rate $\dot{\gamma}$. An apparent viscosity μ_a can be defined as follows:

$$\mu_a = \frac{R}{\dot{\gamma}} \qquad (1.4\text{--}1)$$

When the apparent viscosity μ_a decreases with an increase in shear rate $\dot{\gamma}$ as in Figure (1.4–1) the fluid is said to be pseudoplastic. When

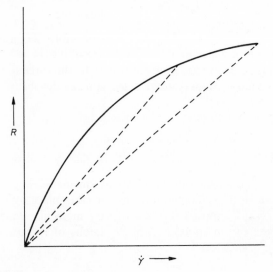

Figure (1.4–1)
Shear stress R against shear rate $\dot{\gamma}$ for a pseudoplastic fluid.

μ_a increases with an increase in $\dot{\gamma}$ as in Figure (1.4–2) the fluid is said to be dilatant.

Another type of non-Newtonian fluid is the Bingham plastic. A plot of R against $\dot{\gamma}$ on Cartesian coordinates for a Bingham plastic shown in Figure (1.4–3) is a straight line having an intercept R_B on the shear stress axis called the yield stress. R_B is the stress which must be exceeded before flow starts. The fluid at rest contains a three dimensional structure of sufficient rigidity to resist any stress less

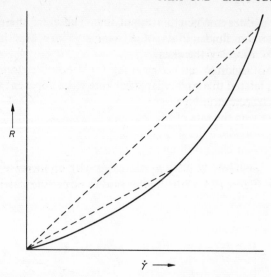

Figure (1.4–2)
Shear stress R against shear rate $\dot{\gamma}$ for a dilatant fluid.

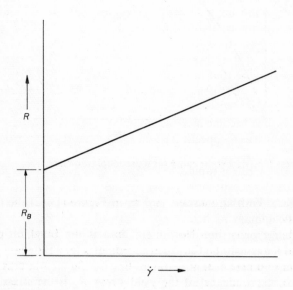

Figure (1.4–3)
Shear stress R against shear rate $\dot{\gamma}$ for a Bingham plastic.

than the yield stress. When this stress is exceeded, the system behaves as a Newtonian fluid under a shear stress $R - R_B$. For Bingham plastics, the slope of the shear stress, shear rate plot is called the coefficient of rigidity.

Pseudoplastics, dilatants and Bingham plastics are examples of time independent non-Newtonian fluids, i.e. the apparent viscosity depends only on the rate of shear at any particular moment and not on the time for which the shear rate is applied.

For a certain class of fluids the apparent viscosity continues to change as a function of the time for which the particular shear rate is applied. These are known as time dependent non-Newtonian materials. Fluids which become more pseudoplastic with time at a constant shear rate are known as thixotropic fluids. Their structure progressively breaks down with time at a constant shear rate. Thixotropy is a reversible process. Eventually, a dynamic equilibrium is reached where the rate of structural breakdown is balanced by the simultaneous rate of reformation. Thus a minimum value of the apparent viscosity is reached at any constant rate of shear. Many fluids show thixotropic behaviour in addition to being pseudoplastic or even dilatant.

Most thixotropic fluids will recover their original viscosity if allowed to stand for a sufficient length of time. Some fluids will revert almost immediately, while others might take several hours.

A plot of shear stress R against shear rate $\dot{\gamma}$ for thixotropic fluids as in Figure (1.4–4) shows a hysteresis effect when the shear rate is changed at regular time intervals. The curve obtained for increasing shear rates does not coincide with the curve for decreasing shear rates.

Fluids which become more dilatant with time at a constant shear rate are known as rheopectic fluids. Small shearing motions facilitate the formation of structure. Above a critical point, breakdown occurs. If the shearing rate is rapid, the structure does not form. In general, the apparent viscosity of rheopectic fluids increases with time to a maximum value at a constant rate of shear. Most rheopectic fluids revert very quickly to their original viscosity if left to stand.

In practice, truly time dependent non-Newtonian fluids are rare.

Another important group of non-Newtonian fluids are viscoelastic fluids. These exhibit both viscous and elastic properties. In a purely elastic solid, the stress corresponding to a given strain is independent of time. In contrast in viscoelastic materials, the stress will gradually dissipate. When viscoelastic materials are extruded

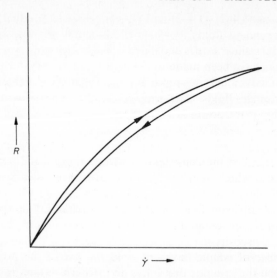

Figure (1.4–4)
Shear stress R against shear rate $\dot{\gamma}$ for a thixotropic fluid exhibiting hysteresis.

through fine perforations the cross-section of the stream may be considerably larger than that of the perforation through which it is extruded. This is the result of partial elastic recovery of the material.

For Newtonian fluids the shear rate $\dot{\gamma}$ is a linear function of the shear stress R given by equation (1.3–5)

$$\dot{\gamma} = \frac{R}{\mu} \qquad (1.3\text{–}5)$$

For non-Newtonian fluids the relationship between $\dot{\gamma}$ and R is more complex and for time independent fluids can be written as

$$\dot{\gamma} = f(R) \qquad (1.4\text{–}2)$$

In terms of velocity in the horizontal x direction, equation (1.4–2) can be rewritten for a point in the z direction in the form

$$\frac{-\mathrm{d}v_x}{\mathrm{d}z} = f(R_{zx}) \qquad (1.4\text{–}3)$$

or for a point in the radial r direction in the form

$$\frac{-\mathrm{d}v_x}{\mathrm{d}r} = f(R_{rx}) \qquad (1.4\text{–}4)$$

Equations (1.4–2), (1.4–3) and (1.4–4) are general equations for time independent fluids in which the function relating the shear stress and the shear rate is undefined.

Attempts have been made to define this function by formulating mathematical models to represent the rheological behaviour of non-Newtonian fluids. The simplest and most commonly used relationship is the power law equation[3]

$$R = K(\dot{\gamma})^n \qquad (1.4–5)$$

where K is called the consistency coefficient and n the power law index. Fluids which obey the above equation are called power law fluids.

For pseudoplastic fluids $n < 1$, and for dilatant fluids $n > 1$. In the case of Newtonian fluids $n = 1$ and K becomes the coefficient of dynamic viscosity μ.

A number of people have criticized the use of the power law equation on the grounds that it has no theoretical basis and that it can be shown to be invalid for unsteady state systems. However, it does give an empirical fit of the data for many fluids which is sufficiently good for engineering purposes.

In terms of velocity in the horizontal x direction, equation (1.4–5) can be rewritten for a point in the z direction in the form

$$R_{zx} = -K\left(\frac{-dv_x}{dz}\right)^{n-1}\frac{dv_x}{dz} \qquad (1.4–6)$$

or for a point in the radial r direction in the form

$$R_{rx} = -K\left(\frac{-dv_x}{dr}\right)^{n-1}\frac{dv_x}{dr} \qquad (1.4–7)$$

Some other mathematical models for time independent non-Newtonian fluids are the Eyring, Ellis and Reiner–Philippoff models.[1]

The Eyring model is a two-parameter model which can be written in the form

$$R_{zx} = A \operatorname{arc\,sinh}\left(-\frac{1}{B}\frac{dv_x}{dz}\right) \qquad (1.4–8)$$

The Ellis model is a three-parameter model which can be written in the form

$$\frac{-dv_x}{dz} = [A + B(R_{zx})^{m-1}]R_{zx} \qquad (1.4–9)$$

The Reiner–Philippoff model[4] is also a three-parameter model. It can be written in the form

$$\frac{-dv_x}{dz} = \frac{R_{zx}}{A + \dfrac{(B - A)}{1 + (R_{zx}/c)^2}} \qquad (1.4\text{--}10)$$

All these rheological mathematical models are empirical curve fitting equations. They are not reliable when used beyond the range of available data. Their parameters are subject to variations by temperature, pressure and other factors.

1.5 Boundary layer

When a fluid flows over a solid surface, its velocity in direct contact with the surface must be zero. Otherwise, the velocity gradient would be infinite at this point and an infinite shear would result at the solid surface. This is readily seen with reference to a fluid which obeys Newton's law of viscosity

$$R = -\mu\frac{dv}{dz} \qquad (1.3\text{--}1)$$

It should be noted with reference to equation (1.3–1) that if the fluid has no viscosity, i.e. $\mu = 0$, then the limitation of no slip at the wall does not apply. Fluids with no viscosity are the so-called ideal fluids which in practice do not exist. However, when dealing with fluids of low viscosity, it is sometimes convenient to consider them as ideal fluids in regions not directly affected by a solid surface.

In the region of immediate proximity to the solid surface, the fluid is directly affected by the solid surface. This region is known as the boundary layer. Prandtl considered a fluid flowing over a solid surface as being divided into two regions: one directly affected by the surface and one unaffected by the surface.

In general, the boundary layer is thin, and since for a Newtonian fluid the shear stress is given by Newton's law of viscosity, its magnitude at a solid surface can be quite high. Outside the boundary layer, velocity gradients and hence shear stresses are small. The concept of the boundary layer leads to a considerable simplification of the theoretical analysis of flow over solid surfaces.

Consider a fluid flowing at a uniform velocity entering a pipe. A boundary layer forms at the pipe wall. This layer grows progressively thicker until it meets the layer from the opposite wall at the axis of the pipe. The thickness of the fully developed boundary layer is thus the radius of the pipe. The type of fluid flow, i.e., streamline or turbulent, in the boundary layer at this point persists in the fully developed flow region. The nature of the fully developed flow in the pipe depends on the density, viscosity, and velocity of the flowing fluid and the diameter of the pipe. At first, the flow in the boundary layer is streamline. If the streamline boundary layer has not filled the pipe after a certain distance from the point of entry, the flow starts to become turbulent. However, even after full turbulence develops, a laminar sublayer remains in the immediate region of the pipe wall.

1.6   Energy relationships and the Bernoulli equation

The total energy of a fluid in motion consists of the following components: internal, potential, pressure, and kinetic energies. Each of these energies may be considered with reference to an arbitrary base level. It is also convenient to make calculations on unit mass of fluid.

Internal energy.   This is the energy associated with the physical state of the fluid, i.e., the energy of the atoms and molecules resulting from their motion and configuration.[5] Internal energy is a function of temperature. It can be written as U per unit mass of fluid.

Potential energy.   This is the energy that a fluid has because of its position in the earth's field of gravity. The work required to raise a unit mass of fluid to a height z above an arbitrarily chosen base level is zg, where g is the gravitational acceleration. This work is equal to the potential energy of unit mass of the fluid above the base level.

Pressure energy.   This is the energy or work required to introduce the fluid into the system without a change in volume. If P is the pressure and V is the volume of a mass m of fluid, then PV/m is the pressure energy per unit mass of fluid. The ratio m/V is the fluid density ρ. Thus the pressure energy per unit mass of fluid can be written as P/ρ.

Kinetic energy.   This is the energy of fluid motion. The kinetic energy of unit mass of the fluid is $v^2/2$, where v is the linear velocity of the fluid relative to some fixed body.

Total energy. The total energy E per unit mass of fluid is given by the equation

$$E = U + zg + \frac{P}{\rho} + \frac{v^2}{2} \qquad (1.6\text{–}1)$$

where each term has the dimensions of force times distance per unit mass, i.e. $(ML/T^2)(L/M)$ or L^2/T^2.

Consider unit mass of fluid flowing from a point 1 to a point 2. Between these two points, let an amount of heat energy Δq be added to the fluid as shown in Figure (1.6–1). Let an amount of work ΔW_1

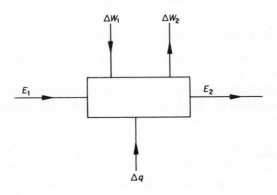

Figure (1.6–1)
Energy balance for a fluid in motion.

be done on the fluid and let the fluid do an amount of work ΔW_2 on its surroundings. An energy balance can be written for unit mass of fluid either as

$$E_1 + \Delta W_1 + \Delta q = E_2 + \Delta W_2$$

or as

$$E_2 - E_1 = \Delta q + \Delta W_1 - \Delta W_2 \qquad (1.6\text{–}2)$$

A flowing fluid is required to do work in order to overcome viscous frictional forces so that in practice the term ΔW_2 is always positive. It is only zero for the theoretical case of an inviscid or ideal fluid having zero viscosity. The work ΔW_1 may be done on the fluid by a pump situated between points 1 and 2.

If the temperature of the flowing fluid remains constant, the internal energy does not change. If no heat is added to the fluid,

$\Delta q = 0$. For these conditions, equations (1.6–1) and (1.6–2) may be combined and written as

$$\left(z_2 g + \frac{P_2}{\rho_2} + \frac{v_2^2}{2}\right) - \left(z_1 g + \frac{P_1}{\rho_1} + \frac{v_1^2}{2}\right) = \Delta W_1 - \Delta W_2 \quad (1.6\text{–}3)$$

For an inviscid fluid and no pump, equation (1.6–3) becomes

$$\left(z_2 g + \frac{P_2}{\rho_2} + \frac{v_2^2}{2}\right) - \left(z_1 g + \frac{P_1}{\rho_1} + \frac{v_1^2}{2}\right) = 0 \qquad (1.6\text{–}4)$$

Equation (1.6–4) is the Bernoulli equation.

Equation (1.6–3) can also be written in the form

$$\left(z_2 + \frac{P_2}{\rho_2 g} + \frac{v_2^2}{2g}\right) - \left(z_1 + \frac{P_1}{\rho_1 g} + \frac{v_1^2}{2g}\right) = \frac{\Delta W_1}{g} - \frac{\Delta W_2}{g} \quad (1.6\text{–}5)$$

where each of the terms has the dimensions of length. In equation (1.6–5), the terms z, $P/(\rho g)$ and $v^2/(2g)$ are known as the potential, pressure and velocity heads, respectively. Equation (1.6–5) can also be written in the form

$$\left(z_2 + \frac{P_2}{\rho_2 g} + \frac{v_2^2}{2g}\right) - \left(z_1 + \frac{P_1}{\rho_1 g} + \frac{v_1^2}{2g}\right) = \Delta h - h_f \quad (1.6\text{–}6)$$

where Δh is the head imparted to the fluid by the pump and h_f is the head loss due to friction. The term Δh is known as the total head of the pump.

Equation (1.6–6) is simply an energy balance written for convenience in terms of length. The various forms of energy are interchangeable and the equation enables the magnitude of these changes to be calculated in a given system.

The Bernoulli equation and the modified Bernoulli equations, equations (1.6–5) and (1.6–6), strictly speaking apply only to the flow of a fluid along a streamline or stream tube. For steady flow in a pipe or tube equation (1.6–6) can be written in a modified form as follows:

$$\left(z_2 + \frac{P_2}{\rho_2 g} + \frac{u_2^2}{2g\alpha_2}\right) - \left(z_1 + \frac{P_1}{\rho_1 g} + \frac{u_1^2}{2g\alpha_1}\right) = \Delta h - h_f \quad (1.6\text{–}7)$$

where u is the mean linear velocity in the pipe or tube and α is a dimensionless correction factor which accounts for the velocity distribution across the pipe or tube. For a pipe of circular cross-section α can be shown to be $\frac{1}{2}$ for laminar flow and approximately 1 for turbulent flow.[2]

For an incompressible fluid of density ρ flowing between points 1 and 2 distance L apart in a pipe of circular cross-section with a constant diameter and no pump, equation (1.6–7) can be written in the form

$$\left(z_1 + \frac{P_1}{\rho g} + \frac{u^2}{2g\alpha}\right) = \left(z_2 + \frac{P_2}{\rho g} + \frac{u^2}{2g\alpha}\right) + h_f \qquad (1.6\text{--}8)$$

Equation (1.6–8) is diagrammatically represented in Figure (1.6–2).

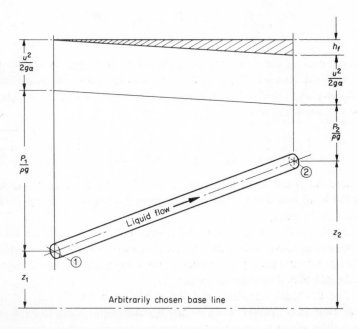

Figure (1.6–2)
Diagrammatic representation of heads in a liquid flowing through a pipe.

In flow systems in which density changes are small, gases can be regarded as incompressible fluids and the same theoretical treatment can be used for both liquids and gases. In the case of high-speed flow or where large temperature differences exist gases must be treated as compressible fluids.

REFERENCES

(1) Bird, R. B., Stewart, W. E., and Lightfoot, E. N., *Transport Phenomena*, p. 12, New York, John Wiley and Sons, Inc., 1960.
(2) Coulson, J. M., and Richardson, J. F., *Chemical Engineering*, Vol. 1, p. 33, New York, The Macmillan Co., 1964.
(3) Ostwald, W., Kollaidzchr., **38**, 261, 1926.
(4) Philippoff, W., Kolloid Z., **71**, 1, 1935.
(5) Smith, J. M., *Introduction to Chemical Engineering Thermodynamics*, p. 6, New York, McGraw-Hill Book Co. Inc., 1949.

2
Flow of incompressible Newtonian fluids in pipes and channels

2.1 Reynolds number and flow patterns in pipes and tubes

The first published work on fluid flow patterns in pipes and tubes was done by Reynolds[8]. He observed the flow patterns of fluids in cylindrical tubes by injecting dye into the moving stream. Reynolds correlated his data by using a dimensionless group N_{RE} later known as the Reynolds number.

$$N_{RE} = \frac{\rho u d_i}{\mu} \qquad (2.1\text{--}1)$$

In equation (2.1–1), ρ is the density, μ the dynamic viscosity, and u the mean linear velocity of the fluid; d_i is the inside diameter of the tube. Any consistent system of units can be used in this equation.

The Reynolds number is also frequently written in the form

$$N_{RE} = \frac{G d_i}{\mu} \qquad (2.1\text{--}2)$$

where $G = \rho u$.

G is commonly known as the mass flow rate in mass per unit area per unit time.

The Reynolds number is sometimes written in the form

$$N_{RE} = \frac{M d_i}{S_i \mu} \qquad (2.1\text{--}3)$$

where M is the flow rate of fluid in mass per unit time and S_i is the cross-sectional flow area in the pipe or tube.

Reynolds found that as he increased the fluid velocity in the tube, the flow pattern changed from laminar to turbulent at a Reynolds number value of about 2100. Later investigators have shown that under certain conditions, e.g. with very smooth conduits, laminar flow can exist at very much higher Reynolds numbers.

2.2 Pressure drop as a function of shear stress at a pipe wall

Consider a fluid flowing with a constant mean linear velocity u through a cylindrical pipe of length L and inside diameter d_i.

A pressure drop ΔP occurs in the pipe because of frictional viscous forces. The latter results in a shear stress R_w over the inside surface of the pipe.

A force balance over the pipe with no slip at the wall gives

$$\Delta P \frac{\pi d_i^2}{4} = R_w \pi d_i L$$

or

$$\Delta P = \frac{4R_w L}{d_i} \tag{2.2–1}$$

which can be written as

$$R_w = \frac{\Delta P}{4L/d_i} \tag{2.2–2}$$

Equations (2.2–1) and (2.2–2) are true, irrespective of the nature of the fluid in the pipe.

2.3 Variation of shear stress in a pipe

Consider a fluid flowing steadily through a cylindrical pipe of length L and inside diameter d_i. Consider a core of fluid of radius r. Let the shear stress at a radial distance r be R_{rx}.

A force balance over the core of fluid in the pipe gives

$$\Delta P \, \pi r^2 = 2\pi r L R_{rx}$$

or

$$\Delta P = \frac{2R_{rx} L}{r} \tag{2.3–1}$$

which can be rewritten either as

$$R_{rx} = \left(\frac{\Delta P}{L}\right)\frac{r}{2} \qquad (2.3\text{–}2)$$

or as

$$r = \frac{2LR_{rx}}{\Delta P} \qquad (2.3\text{–}3)$$

Combine equations (2.3–3) and (2.2–1) to give

$$r = \frac{R_{rx}d_i}{2R_w} \qquad (2.3\text{–}4)$$

It is seen from equation (2.3–4) that the shear stress R_{rx} is zero at the centre of the pipe and has a maximum value R_w at the pipe wall.

The mean shear stress R_m over the fluid in the pipe is given by the equation

$$R_m = \frac{\int_0^{d_i/2} R_{rx}\,dr}{\int_0^{d_i/2}\,dr} \qquad (2.3\text{–}5)$$

Combine equation (2.3–4) with equation (2.3–5) and carry out the integration to give

$$R_m = \frac{R_w}{2} \qquad (2.3\text{–}6)$$

Equations (2.3–1), (2.3–2), (2.3–3), (2.3–4), (2.3–5) and (2.3–6) are true irrespective of the nature of the fluid in the pipe.

2.4 Friction factor and pressure drop as a function of Reynolds number in a pipe

Equation (2.2–1) gives the pressure drop in a cylindrical pipe in terms of the shear stress R_w over the inside surface of the pipe.

$$\Delta P = \frac{4R_w L}{d_i} \qquad (2.2\text{–}1)$$

Rewrite equation (2.2–1) in the form

$$\Delta P = 8\left(\frac{R_w}{\rho u^2}\right)\left(\frac{L}{d_i}\right)\frac{\rho u^2}{2} \qquad (2.4\text{–}1)$$

where the term in the first brackets is the dimensionless basic friction factor j_f.

$$j_f = \frac{R_w}{\rho u^2} \tag{2.4-2}$$

Thus equation (2.4–1) can be written as

$$\Delta P = 8j_f\left(\frac{L}{d_i}\right)\frac{\rho u^2}{2} \tag{2.4-3}$$

Equation (2.4–3) is true irrespective of the flow patterns in the pipe or the nature of the fluid.

The basic friction factor j_f is half the Fanning friction factor f. In terms of f, equation (2.4–3) can be written as

$$\Delta P = 4f\left(\frac{L}{d_i}\right)\frac{\rho u^2}{2} \tag{2.4-4}$$

For laminar flow

$$j_f = \frac{8}{N_{RE}} \tag{2.4-5}$$

Equation (2.4–5) can be substituted into equation (2.4–3) to give the Hagen–Poiseuille equation for steady state laminar flow of Newtonian fluids in pipes and tubes.

$$u = \left(\frac{\Delta P}{L}\right)\frac{d_i^2}{32\mu} \tag{2.4-6}$$

Equation (2.4–6) can also be written in the form

$$\frac{\Delta P}{4L/d_i} = \mu\left(\frac{8u}{d_i}\right) \tag{2.4-7}$$

where the term $8u/d_i$ is known as the flow characteristic.

Since equation (2.4–7) can be combined with equation (2.2–2) to read

$$R_w = \mu\left(\frac{8u}{d_i}\right) \tag{2.4-8}$$

or

$$\frac{8u}{d_i} = \frac{R_w}{\mu} \tag{2.4-9}$$

the flow characteristic $8u/d_i$ must equal the shear rate $\dot{\gamma}_w$ at the pipe wall for a Newtonian fluid in laminar flow, since from Newton's law of viscosity

$$R_w = \mu\dot{\gamma}_w \tag{2.4-10}$$

or

$$\dot{\gamma}_w = \frac{R_w}{\mu} \tag{2.4-11}$$

For the turbulent flow of Newtonian fluids in smooth cylindrical pipes and tubes, j_f is approximately related to N_{RE} by the equation

$$j_f = \frac{0.0396}{N_{RE}^{0.25}} \tag{2.4-12}$$

Equation (2.4–12) is an empirical equation which holds for Reynolds numbers up to 10^5.

In terms of the Fanning friction factor f, equation (2.4–12) is known as the Blasius equation[3] which is usually written as

$$f = \frac{0.079}{N_{RE}^{0.25}} \tag{2.4-13}$$

An alternative to equation (2.4–13) for the turbulent flow of Newtonian fluids in smooth cylindrical pipes is the von Karman equation which is normally written in terms of f as follows:

$$\frac{1}{f^{\frac{1}{2}}} = 4.0 \log(N_{RE}f^{\frac{1}{2}}) - 0.40 \tag{2.4-14}$$

For pipes which are not smooth, the friction factors f and j_f for turbulent flow are a function not only of N_{RE} but also of a dimensionless roughness factor ε/d_i where ε is a linear quantity representing the roughness of the pipe surface. Values of ε for various kinds of pipes are given in Table (2.4–1).

TABLE (2.4–1)

	absolute roughness ε in m
drawn tubing	0.0000015
commercial steel and wrought iron	0.000045
asphalted cast iron	0.00012
galvanized iron	0.00015
cast iron	0.00026
wood stave	0.00018–0.00092
concrete	0.00030–0.0030
riveted steel	0.00092–0.0092

For rough pipes, f for turbulent flow is related to ε/d_i by another von Karman equation

$$\frac{1}{f^{\frac{1}{2}}} = 4.06 \log\left(\frac{d_i}{\varepsilon}\right) + 2.16 \qquad (2.4\text{--}15)$$

Friction factor data are conventionally given in log-log plots of friction factor against Reynolds number such as the j_f against N_{RE} plot in Figure (2.4–1).

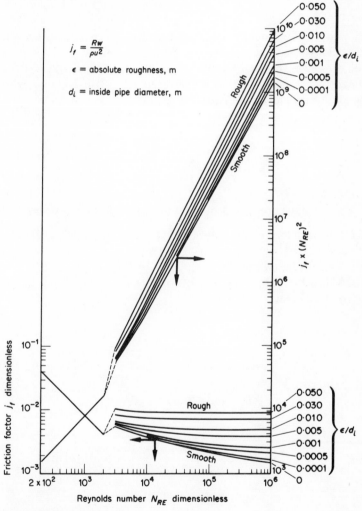

Figure (2.4–1)
Friction factor j_f against Reynolds number plot.

For turbulent flow, it is not possible to determine directly the fluid flow rate through a pipe from a j_f against N_{RE} plot for a known pressure drop. However, it is possible to do this from a $j_f(N_{RE})^2$ against N_{RE} plot also shown in Figure (2.4–1).

It can readily be shown by combining equations (2.1–1), (2.2–1), and (2.4–2) that

$$j_f(N_{RE})^2 = \frac{\Delta P \, \rho d_i^3}{4L\mu^2} \qquad (2.4\text{–}16)$$

Equation (2.4–16) does not contain the mean linear velocity u of the fluid. This can be determined as follows. Calculate $j_f(N_{RE})^2$ from equation (2.4–16) from known values of ΔP, ρ, d_i, L and μ. Read the corresponding value of N_{RE} from Figure (2.4–1) for a known value of ε/d_i. Calculate u from equation (2.1–1) from known values of ρ, d_i and μ.

$$N_{RE} = \frac{\rho u d_i}{\mu} \qquad (2.1\text{–}1)$$

Example (2.4–1)
   Calculate the pressure drop in N/m^2 for the following system:

(1) A 30.48 m long commercial steel pipe of inside diameter 0.0526 m and a pipe roughness $\varepsilon = 0.000045$ m.
(2) A steady liquid transfer rate of 9.085 m^3/h.
(3) A liquid dynamic viscosity of 0.01 N s/m^2 and a liquid density of 1200 kg/m^3.

Calculations:

$$\text{mean linear velocity } u = \frac{Q}{\pi d_i^2/4} \qquad (2.7\text{–}8)$$

$$\frac{\pi d_i^2}{4} = \frac{(3.142)(0.0526 \text{ m})^2}{4} = 0.002173 \text{ m}^2$$

$$Q = 9.085 \text{ m}^3/\text{h} = 9.085 \frac{\text{m}^3}{\text{h}} \left(\frac{1 \text{ h}}{3600 \text{ s}} \right) = 0.002524 \text{ m}^3/\text{s}$$

$$u = \frac{0.002524 \text{ m}^3/\text{s}}{0.002173 \text{ m}^2} = 1.160 \text{ m/s}$$

$$\text{Reynolds number } N_{RE} = \frac{\rho u d_i}{\mu} \tag{2.1-1}$$

$\rho = 1200 \text{ kg/m}^3$

$u = 1.160 \text{ m/s}$

$d_i = 0.0526 \text{ m}$

$\mu = 0.01 \text{ N s/m}^2 = 0.01 \text{ kg/(s m)}$

$$N_{RE} = \frac{(1200 \text{ kg/m}^3)(1.160 \text{ m/s})(0.0526 \text{ m})}{0.01 \text{ kg/(s m)}} = 7322$$

pipe roughness $\varepsilon = 0.000045 \text{ m}$

$$d_i = 0.0526 \text{ m}$$

$$\text{roughness factor } \frac{\varepsilon}{d_i} = \frac{0.000045 \text{ m}}{0.0526 \text{ m}} = 0.000856$$

from j_f against N_{RE} graph in Figure (2.4–1), $j_f = 0.0042$ for $N_{RE} = 7322$ and $\varepsilon/d_i = 0.000856$

$$\frac{L}{d_i} = \frac{30.48 \text{ m}}{0.0526 \text{ m}} = 579.5$$

$$\frac{\rho u^2}{2} = \frac{(1200 \text{ kg/m}^3)(1.160 \text{ m/s})^2}{2}$$

$$= 807.4 \text{ kg/(s}^2 \text{ m)} = 807.4 \text{ N/m}^2$$

$$\text{pressure drop } \Delta P = 8 j_f \left(\frac{L}{d_i} \right) \frac{\rho u^2}{2} \tag{2.4-3}$$

$$= 8(0.0042)(579.5)(807.4 \text{ N/m}^2)$$

$$= 15\,720 \text{ N/m}^2$$

Example (2.4–2)

Estimate the steady mean linear velocity in m/s inside a pipe for the following system:

(1) A 30.48 m long commercial steel pipe of inside diameter 0.0526 m and a pipe roughness $\varepsilon = 0.000045$ m.

(2) A pressure drop in the pipe of 15 720 N/m².

(3) A liquid dynamic viscosity of 0.01 N s/m² and a liquid density of 1200 kg/m³.

Calculations:

$$j_f(N_{RE})^2 = \frac{\Delta P \, \rho d_i^3}{4L\mu^2} \qquad (2.4\text{--}16)$$

$$\Delta P = 15\,720 \text{ N/m}^2 = 15\,720 \text{ kg/(s}^2 \text{ m)}$$

$$\rho = 1200 \text{ kg/m}^3$$

$$d_i = 0.0526 \text{ m}$$

$$L = 30.48 \text{ m}$$

$$\mu = 0.01 \text{ N s/m}^2 = 0.01 \text{ kg/(s m)}$$

$$\frac{\Delta P \, \rho d_i^3}{4L\mu^2} = \frac{[15\,720 \text{ kg/(s}^2 \text{ m)}](1200 \text{ kg/m}^3)(0.0526 \text{ m})^3}{(4)(30.48 \text{ m})[0.01 \text{ kg/(s m)}]^2}$$

$$= 2.252 \times 10^5$$

$$\text{roughness factor } \frac{\varepsilon}{d_i} = \frac{0.000045 \text{ m}}{0.0526 \text{ m}} = 0.000856$$

from $j_f(N_{RE})^2$ against N_{RE} graph in Figure (2.4–1), $N_{RE} = 7200$ for $j_f(N_{RE})^2 = 2.252 \times 10^5$ and $\varepsilon/d_i = 0.000856$

$$\text{mean linear velocity } u = \frac{N_{RE}\mu}{\rho d_i}$$

$$= \frac{(7200)[0.01 \text{ kg/(s m)}]}{(1200 \text{ kg/m}^3)(0.0526 \text{ m})}$$

$$= 1.141 \text{ m/s}$$

The slight diffference between this and the mean linear velocity in Example (2.4–1) is due to error in reading the graph in Figure (2.4–1).

2.5   Pressure drop in fittings and curved pipes

So far, only the pressure drop in straight lengths of pipe of circular cross-section has been discussed. The pressure drop in pipelines containing valves and fittings can be calculated from equation (2.4–3) written in the modified form

$$\Delta P = 8 j_f \left(\frac{\Sigma L_e}{d_i} \right) \frac{\rho u^2}{2} \tag{2.5–1}$$

where ΣL_e is the sum of the equivalent lengths of the components in the pipeline. Equivalent lengths of various valves and fittings are readily available[4].

Pipe entrance and exit pressure losses should also be calculated and added to get the overall pressure drop.

The loss in pressure due to sudden expansion from a diameter d_{i1} to a larger diameter d_{i2} is given by the equation

$$\Delta P_e = \frac{\rho(u_1 - u_2)^2}{2} = \frac{\rho u_1^2}{2} \left[1 - \left(\frac{d_{i1}}{d_{i2}} \right)^2 \right]^2 \tag{2.5–2}$$

where u_1 and u_2 are the mean linear velocities in the smaller entrance pipe and the larger exit pipe respectively.

The loss in pressure due to sudden contraction from a diameter d_{i1} to a smaller diameter d_{i2} is given by the equation

$$\Delta P_c = K \left(\frac{\rho u_2^2}{2} \right) \tag{2.5–3}$$

where

$$K = 0.4 \left[1.25 - \left(\frac{d_{i2}}{d_{i1}} \right)^2 \right] \quad \text{when} \quad \frac{d_{i2}^2}{d_{i1}^2} < 0.715$$

and

$$K = 0.75 \left[1.0 - \left(\frac{d_{i2}}{d_{i1}} \right)^2 \right] \quad \text{when} \quad \frac{d_{i2}^2}{d_{i1}^2} > 0.715,$$

u_2 is the mean linear velocity in the smaller exit pipe.

A number of equations have been proposed for use in the calculation of pressure drops in coils of constant curvature.[9] The latter are known as helices. For laminar flow, Kubair and Kuloor[6] gave an equation for the Reynolds number range 170 to the critical value. In terms of the basic friction factor their equation can be written as

$$(j_f)_c = \frac{8[2.8 + 12(d_i/D_c)]}{N_{RE}^{1.15}}$$ (2.5–4)

where d_i and D_c are the tube and coil diameters respectively. The critical Reynolds number may be defined as the highest Reynolds number for which the flow in a helix is still definitely in the viscous or laminar region. The critical Reynolds number can be calculated from the equation

$$(N_{RE})_{\text{critical}} = 2100\left[1 + 12\left(\frac{d_i}{D_c}\right)^{0.5}\right]$$ (2.5–5)

For turbulent flow, White[10] gave an equation for the Reynolds number range 15 000 to 100 000. In terms of the basic friction factor, White's equation can be written as

$$(j_f)_c = \frac{0.04}{N_{RE}^{0.25}} + 0.006\left(\frac{d_i}{D_c}\right)^{0.5}$$ (2.5–6)

For pipes of noncircular cross-section an equivalent diameter should be used in place of d_i in equations (2.4–3), (2.4–4) and (2.5–1).

2.6   Equivalent diameter for noncircular pipes

For shapes other than circular, the inside diameter d_i in equations (2.4–3), (2.4–4) and (2.5–1) can be replaced by an equivalent diameter d_e defined as four times the cross-sectional flow area S_i divided by the appropriate flow perimeter.

For a circular cross-section

$$d_e = \frac{4(\pi d_i^2/4)}{\pi d_i} = d_i$$ (2.6–1)

For an annulus the equivalent diameter d_{ep} for pressure drop differs from the equivalent diameter d_{eq} for heat transfer between the annulus and the inner conduit since the appropriate flow perimeters differ in the two cases.[5]

Consider a pipe of circular cross-section with an inside and an outside diameter of d_i and d_o respectively. Let this pipe be placed symmetrically inside a larger pipe having an inside diameter of D_i and let a fluid flow through the annulus.

Since the shear stress resisting the flow of fluid acts on both walls of the annulus, the appropriate flow perimeter required to calculate the equivalent diameter of the annulus d_{ep} for pressure drop is $(\pi D_i + \pi d_o)$.

Therefore

$$d_{ep} = \frac{4[(\pi D_i^2/4) - (\pi d_o^2/4)]}{\pi D_i + \pi d_o}$$

$$= D_i - d_o \qquad (2.6\text{--}2)$$

However since in heat transfer between the annulus and the inner pipe only one wall is involved, the appropriate flow perimeter is πd_o. Therefore the equivalent diameter of the annulus d_{eq} for heat transfer is

$$d_{eq} = \frac{4[(\pi D_i^2/4) - (\pi d_o^2/4)]}{\pi d_o}$$

$$= \frac{(D_i^2 - d_o^2)}{d_o} \qquad (2.6\text{--}3)$$

2.7 Velocity distribution for laminar flow in a pipe

It has already been shown that the shear stress R_{rx} at a radial distance r in a pipe is given by equation (2.3–2).

$$R_{rx} = \left(\frac{\Delta P}{L}\right)\frac{r}{2} \qquad (2.3\text{--}2)$$

For a Newtonian fluid in laminar flow, R_{rx} is related to the velocity gradient $-dv_x/dr$ at a radial distance r by equation (1.3–3)

$$R_{rx} = -\frac{\mu\, dv_x}{dr} \qquad (1.3\text{--}3)$$

Combine equations (2.3–2) and (1.3–3) to give

$$\frac{dv_x}{dr} = -\left(\frac{\Delta P}{L}\right)\frac{r}{2\mu} \qquad (2.7\text{--}1)$$

and integrate to

$$v_x = -\left(\frac{\Delta P}{L}\right)\frac{r^2}{4\mu} + C \qquad (2.7\text{--}2)$$

Since the linear velocity $v_x = 0$ at $r = d_i/2$, the constant

$$C = \left(\frac{\Delta P}{L}\right)\frac{d_i^2}{16\mu}$$

and equation (2.7–2) can be written

$$v_x = \left(\frac{\Delta P}{L}\right)\left(\frac{d_i^2}{16\mu}\right)\left[1 - \left(\frac{2r}{d_i}\right)^2\right] \qquad (2.7\text{–}3)$$

Equation (2.7–3) gives the point linear velocity v_x at any radial distance r in the pipe. Thus the velocity profile is parabolic for steady laminar flow of a Newtonian fluid through a pipe of circular cross-section.

The point linear velocity v_x is a maximum at the centre of the pipe, i.e., when $r = 0$. Therefore

$$v_{max} = \left(\frac{\Delta P}{L}\right)\frac{d_i^2}{16\mu} \qquad (2.7\text{–}4)$$

Consider an annular element of fluid of width dr at a radial distance r from the centre of the pipe as shown in Figure (2.7–1).

The volumetric flow rate through the annular element is

$$dQ = 2\pi r\, dr v_x \qquad (2.7\text{–}5)$$

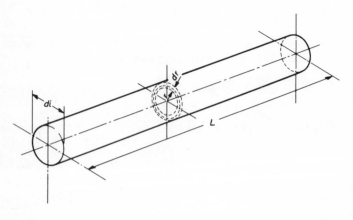

Figure (2.7–1)
Annulus of flow in a pipe.

Thus the total volumetric flow rate through a pipe of diameter d_i is

$$Q = \int_0^{d_i/2} 2\pi r v_x \, dr \tag{2.7-6}$$

Substitute equation (2.7–3) into equation (2.7–6) and integrate to give

$$Q = \left(\frac{\Delta P}{L}\right)\frac{\pi d_i^4}{128\mu} \tag{2.7-7}$$

which is the volumetric flow rate of a Newtonian fluid in steady laminar flow through a pipe.

The mean linear velocity u in the pipe is

$$u = \frac{Q}{\pi d_i^2/4} \tag{2.7-8}$$

Combine equations (2.7–7) and (2.7–8) to give

$$u = \left(\frac{\Delta P}{L}\right)\frac{d_i^2}{32\mu} \tag{2.4-6}$$

which is the Hagen–Poiseuille equation for steady laminar flow already derived in Section 2.4 assuming equation (2.4–5)

$$j_f = \frac{8}{N_{RE}} \tag{2.4-5}$$

Thus the derivation of the Hagen–Poiseuille equation in this section is a theoretical proof of equation (2.4–5).

Equation (2.7–3) which gives the laminar flow velocity profile in a pipe, can also be written either as

$$\frac{v_x}{v_{max}} = 1 - \left(\frac{2r}{d_i}\right)^2 \tag{2.7-9}$$

or as

$$v_x = 2u\left[1 - \left(\frac{2r}{d_i}\right)^2\right] \tag{2.7-10}$$

For the steady laminar flow of a Newtonian fluid in a pipe of circular cross-section, the mean linear velocity u is related to the maximum point linear velocity v_{max} by the equation

$$\frac{u}{v_{max}} = 0.5 \tag{2.7-11}$$

Again consider the annular element of fluid in Figure (2.7–1). The rate of flow of kinetic energy through the annular element is

$$d(K.E) = \rho \, dQ \, \frac{v_x^2}{2} \qquad (2.7\text{--}12)$$

Thus the total rate of flow of kinetic energy through a pipe of diameter d_i is

$$K.E = \int_0^{d_i/2} \frac{\rho v_x^2}{2} \, dQ \qquad (2.7\text{--}13)$$

Substitute equations (2.7–3) and (2.7–5) into equation (2.7–13) and integrate to give

$$K.E = \left(\frac{\Delta P}{4\mu L}\right)^3 \frac{\rho \pi d_i^8}{2048} \qquad (2.7\text{--}14)$$

The flow rate through the pipe in mass per unit time is

$$M = \rho Q \qquad (2.7\text{--}15)$$

Substitute equation (2.7–7) into equation (2.7–15) to give

$$M = \left(\frac{\Delta P}{L}\right) \frac{\rho \pi d_i^4}{128\mu} \qquad (2.7\text{--}16)$$

The kinetic energy per unit mass

$$\frac{K.E}{M} = \left(\frac{\Delta P}{4\mu L}\right)^3 \frac{\rho \pi d_i^8}{2048} \bigg/ \left[\left(\frac{\Delta P}{L}\right) \frac{\rho \pi d_i^4}{128\mu}\right] \qquad (2.7\text{--}17)$$

which can also be written

$$\frac{K.E}{M} = \left[\left(\frac{\Delta P}{L}\right) \frac{d_i^2}{32\mu}\right]^2 = u^2 \qquad (2.7\text{--}18)$$

Thus the kinetic energy per unit mass of a Newtonian fluid in steady laminar flow through a pipe of circular cross–section is u^2. In terms of head this is u^2/g. Therefore for laminar flow, $\alpha = \frac{1}{2}$ in equation (1.6–8).

2.8   Velocity distribution for turbulent flow in a pipe

The theory for the turbulent flow of fluids through pipes is far less developed than that for laminar flow.

The velocity profile equation for steady turbulent flow of a Newtonian fluid through a pipe of circular cross-section corresponding to equation (2.7–9) for laminar flow is

$$\frac{v_x}{v_{max}} = \left(1 - \frac{2r}{d_i}\right)^{1/7} \tag{2.8–1}$$

Equation (2.8–1) is an empirical equation known as the one-seventh power velocity distribution equation for turbulent flow. It fits the experimentally determined velocity distribution data of Nikuradse with a fair degree of accuracy. Equation (2.8–1) is not valid for fluid which is in immediate proximity to the pipe wall.

Equation (2.8–1) is also commonly written in the form

$$\frac{v_x}{v_{max}} = \left(\frac{2z}{d_i}\right)^{1/7} \tag{2.8–2}$$

where z is the distance from the pipe wall.

$$z = \frac{d_i}{2} - r \tag{2.8–3}$$

For the steady turbulent flow of a Newtonian fluid in a pipe of circular cross-section, the mean linear velocity u is related to the maximum point linear velocity v_{max} by the equation

$$\frac{u}{v_{max}} = 0.8 \tag{2.8–4}$$

Thus the turbulent flow velocity profile curve is flatter than the corresponding laminar flow curve.

Equation (2.8–4) can be proved as follows. Consider an annular element of fluid of width dr at a radial distance r from the centre of the pipe as shown in Figure (2.7–1).

The volumetric flow rate through the annular element is

$$dQ = 2\pi r \, dr v_x \tag{2.8–5}$$

which in terms of equation (2.8–2) and z the distance from the pipe wall can be written as

$$dQ = 2\pi v_{max}\left(\frac{d_i}{2} - z\right)\left(\frac{2z}{d_i}\right)^{1/7}(-dz) \tag{2.8–6}$$

Integrate equation (2.7–6) between the limits $z = 0$ and $z = d_i/2$ to give Q the total volumetric flow rate through a pipe of diameter d_i.

$$Q = \frac{49}{60}\left(\frac{\pi d_i^2}{4}\right)v_{\max} \qquad (2.8\text{–}7)$$

The mean linear velocity

$$u = \frac{Q}{\pi d_i^2/4} \qquad (2.7\text{–}8)$$

Combine equations (2.7–8) and (2.8–7) and write

$$u = \frac{49}{60}v_{\max} \qquad (2.8\text{–}8)$$

which can also be approximately written as

$$\frac{u}{v_{\max}} = 0.8 \qquad (2.8\text{–}4)$$

It has already been shown that the total rate of flow of kinetic energy through a pipe of diameter d_i is

$$K.E = \int_0^{d_i/2} \frac{\rho v_x^2}{2}\, dQ \qquad (2.7\text{–}13)$$

Substitute equations (2.8–2) and (2.8–6) into equation (2.7–13) and integrate to give

$$K.E = \frac{49}{170}\left(\frac{\pi d_i^2}{4}\right)\rho v_{\max}^3 \qquad (2.8\text{–}9)$$

The flow rate through the pipe in mass per unit time is

$$M = \rho Q \qquad (2.7\text{–}15)$$

Substitute equation (2.8–7) into equation (2.7–15) to give

$$M = \frac{49}{60}\left(\frac{\pi d_i^2}{4}\right)\rho v_{\max} \qquad (2.8\text{–}10)$$

Divide equation (2.8–9) by equation (2.8–10) to give the kinetic energy per unit mass

$$\frac{K.E}{M} = 0.353 v_{\max}^2 \qquad (2.8\text{–}11)$$

Equation (2.8–11) can be combined with equation (2.8–4) and written in the approximate form

$$\frac{K.E}{M} = \frac{u^2}{2} \qquad (2.8\text{–}12)$$

Thus the kinetic energy per unit mass of a Newtonian fluid in steady turbulent flow through a pipe of circular cross-section is $u^2/2$. In terms of head this is $u^2/(2g)$. Therefore for turbulent flow, $\alpha = 1$ in equation (1.6–8).

2.9   Universal velocity distribution for turbulent flow in a pipe

Consider a fully developed turbulent flow through a pipe of circular cross-section with a thin laminar sublayer immediately adjacent to the wall. In the laminar sublayer, the shear stress R is given by Newton's law of viscosity which can be written in this case as

$$R = \frac{\mu \, dv_x}{dz} \qquad (2.9\text{–}1)$$

since the velocity gradient dv_x/dz is positive.

Assume that R is constant throughout the laminar sublayer and integrate equation (2.9–1) to give

$$Rz = \mu v_x + C \qquad (2.9\text{–}2)$$

where C is a constant. Since the linear velocity $v_x = 0$ at $z = 0$, $C = 0$. Therefore, equation (2.9–2) can be rewritten as

$$v_x = \frac{Rz}{\mu} = \frac{Rz}{\rho\eta} \qquad (2.9\text{–}3)$$

where η is the kinematic viscosity.

The term R/ρ in equation (2.9–3) is a constant and has the dimensions of velocity squared.

$$v^* = \sqrt{\frac{R}{\rho}} \qquad (2.9\text{–}4)$$

where v^* is commonly known as the friction velocity or the shear stress velocity.

Combine equations (2.9–3) and (2.9–4) and write

$$\frac{v_x}{v^*} = \frac{v^* z}{\eta} = \frac{\rho v^* z}{\mu} \qquad (2.9\text{–}5)$$

Equation (2.9–5) can also be written as

$$v^+ = z^+ \tag{2.9–6}$$

where v^+ is defined as a dimensionless velocity v_x/v^* and z^+ is defined as a dimensionless distance v^*z/η; z^+ has the form of a Reynolds number. Equation (2.9–6) fits the experimental data in the range $0 < z^+ < 5$. In the laminar sublayer, the velocity increases linearly with distance from the wall in contrast to the parabolic velocity profile for fully developed laminar flow in a pipe of circular cross-section.

In a turbulent fluid, the shear stress R is given by equation (1.3–8). The eddy viscous diffusivity η_e is much greater than the molecular diffusivity η and for the turbulent flow region in the pipe, R can be written as

$$R = \eta_e \rho \frac{dv_x}{dz} \tag{2.9–7}$$

where the velocity gradient dv_x/dz is positive.

Prandtl[7] assumed that in turbulent flow, eddies move about in a similar manner to molecules in a gas. Prandtl defined a mixing length L for turbulent flow analogous to the mean free path in the kinetic theory of gases.

It can be shown[2] that

$$\eta_e = L^2 \frac{dv_x}{dz} \tag{2.9–8}$$

Substitute equation (2.9–8) into equation (2.9–7) to give

$$R = \rho L^2 \left(\frac{dv_x}{dz}\right)^2 \tag{2.9–9}$$

Prandtl assumed that L was proportional to z, the distance away from the solid wall. This is reasonable since L must be zero at the wall.

Write

$$L = Kz \tag{2.9–10}$$

where K is a proportionality constant.

Substitute equation (2.9–10) into equation (2.9–9) to give

$$\frac{R}{\rho} = K^2 z^2 \left(\frac{dv_x}{dz}\right)^2 \tag{2.9–11}$$

Combine equations (2.9–4) and (2.9–11) and write

$$v^* = Kz\frac{dv_x}{dz} \tag{2.9–12}$$

Integrate equation (2.9–12) to give

$$v_x = v^*\left(\frac{1}{K}\ln z + C_1\right) \tag{2.9–13}$$

where C_1 is a constant.

Rewrite equation (2.9–13) in the modified form

$$v_x = v^*\left[\frac{1}{K}\ln\left(\frac{\rho v^* z}{\mu}\right) + C_2\right] \tag{2.9–14}$$

where C_2 is another constant. Equation (2.9–14) cannot be applied near the wall, since it gives $v_x = -\infty$ instead of $v_x = 0$ at $z = 0$.

Rewrite equation (2.9–14) in terms of $v^+ = v_x/v^*$ and $z^+ = \rho v^* z/\mu$ to give

$$v^+ = \frac{1}{K}\ln z^+ + C \tag{2.9–15}$$

where C is a constant.

Equation (2.9–15) fits the experimental data for turbulent flow in smooth pipes of circular cross-section for $z^+ > 30$ when written in the form

$$v^+ = 2.5\ln z^+ + 5.5 \tag{2.9–16}$$

For the buffer region $5 < z^+ < 30$ the appropriate equation is

$$v^+ = 5.0\ln z^+ - 3.05 \tag{2.9–17}$$

Equations (2.9–6), (2.9–16) and (2.9–17) enable the velocity distribution to be calculated for steady turbulent flow in a pipe of circular cross-section. These equations are only approximate and lead to discontinuities at $z^+ = 5$ and $z^+ = 30$.

2.10 Flow characteristic as a function of velocity gradient in a pipe

The total volumetric flow rate through a pipe of circular cross-section is given by the equation

$$Q = \int_0^{d_i/2} 2\pi r v_x\, dr \tag{2.7–6}$$

Equation (2.7–6) can be integrated by parts.

Since

$$d(zy) = z\,dy + y\,dz$$

$$zy = \int z\,dy + \int y\,dz$$

and

$$\int y\,dz = zy - \int z\,dy \qquad (2.10\text{--}1)$$

Let $z = \pi r^2$ and $y = v_x$.

Therefore $dz = 2\pi r\,dr$, $dy = dv_x$ and

$$\int y\,dz = \int 2\pi r v_x\,dr$$

Equation (2.7–6) can be written in terms of equation (2.10–1) as

$$Q = \int_0^{d_i/2} 2\pi r v_x\,dr = (\pi r^2 v_x)_0^{d_i/2} - \int_0^{d_i/2} \pi r^2 \left(\frac{dv_x}{dr}\right) dr \quad (2.10\text{--}2)$$

Since at the pipe wall $r = d_i/2$ and the linear velocity $v_x = 0$ assuming no slip

$$(\pi r^2 v_x)_0^{d_i/2} = 0$$

Thus the volumetric flow rate Q can be written in terms of the velocity gradient $-dv_x/dr$ as follows:

$$Q = \int_0^{d_i/2} \pi r^2 \left(\frac{-dv_x}{dr}\right) dr \qquad (2.10\text{--}3)$$

The mean linear velocity u is related to Q by the equation

$$u = \frac{Q}{\pi d_i^2/4} \qquad (2.7\text{--}8)$$

Combine equations (2.10–3) and (2.7–8) and write

$$\frac{8u}{d_i} = \frac{32}{d_i^3} \int_0^{d_i/2} r^2 \left(\frac{-dv_x}{dr}\right) dr \qquad (2.10\text{--}4)$$

where $8u/d_i$ is the flow characteristic. Equation (2.10–4) gives $8u/d_i$ in terms of the velocity gradient at a radial point in the pipe $-dv_x/dr$.

2.11   Flow in open channels

Consider a liquid flowing in an open channel of uniform cross-section under the influence of gravity. The liquid has a free surface subjected only to atmospheric pressure. If the flow is steady, the depth of the liquid is uniform and the hydraulic slope of the free liquid surface is parallel with the slope of the channel bed. Consider a length ΔL in Figure (2.11–1) in which the frictional head loss is Δh_f.

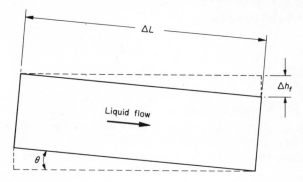

Figure (2.11–1)
Flow in an open channel.

Let the channel slope at a small angle θ to the horizontal. The slope of the channel flow $s = \sin \theta$.

The frictional head loss is given by the equations

$$\Delta h_f = \Delta L s \tag{2.11–1}$$

and

$$\Delta h_f = 8 j_f \left(\frac{\Delta L}{d_e}\right)\frac{u^2}{2g} \tag{2.11–2}$$

Equation (2.11–2) is another way of writing equation (2.4–3) where, in this case, the pressure drop is expressed in height of fluid instead of in force per unit area. In equation (2.11–2), d_e is the equivalent diameter defined as four times the cross-sectional flow area divided by the appropriate flow perimeter, j_f is the dimensionless friction factor for flow in an open channel and u is the mean linear velocity.

Combine equations (2.11–1) and (2.11–2) and solve for u to give

$$u = \sqrt{\frac{g}{j_f}}\sqrt{\frac{d_e s}{4}} \tag{2.11–3}$$

where the mean linear velocity u is proportional to the square root of the channel slope s.

Equation (2.11–3) is frequently written in the form

$$u = C \sqrt{\frac{d_e s}{4}} \tag{2.11–4}$$

which is known as the Chezy formula.

The Chezy coefficient

$$C = \sqrt{\frac{g}{j_f}} \tag{2.11–5}$$

Manning and others[1] gave values of C for various types of surface roughness. A typical value for C when water flows in a concrete channel is $100\ \mathrm{m}^{\frac{1}{2}}/\mathrm{s}$. In general, liquids such as water which commonly flow in open channels have a low viscosity and the flow is almost always turbulent.

REFERENCES

(1) Barna, P. S., *Fluid Mechanics for Engineers*, p. 85, London, Butterworths, 1969.
(2) Bennet, C. O., and Myers, J. E., *Momentum, Heat and Mass Transfer*, p. 131, New York, McGraw-Hill Book Co. Inc., 1962.
(3) Blasius, H., Forsch, Ver. Deut. Ing., 131 (1913).
(4) Holland, F. A., and Chapman, F. S., *Pumping of Liquids*, p. 79, New York, Reinhold Publishing Corporation, 1966.
(5) Holland, F. A., Moores, R. M., Watson, F. A., and Wilkinson, J. K., *Heat Transfer*, p. 461, London, Heinemann Educational Books Ltd., 1970.
(6) Kubair, V., and Kuloor, N. R., Indian Journal of Technology, **3**, 5 (1965).
(7) Prandtl, L. Z., Angew. Math. Mech., **5**, 136 (1925).
(8) Reynolds, O., *Proceedings Royal Society* (London), A, **174**, 935 (1883).
(9) Srinivasan, P. S., Nandapurkar, S. S., and Holland, F. A., The Chemical Engineer, No. 218 (1968).
(10) White, C. M., *Transactions Institution of Chemical Engineers*, **10**, 66 (1932).

3
Flow of incompressible non-Newtonian fluids in pipes

3.1 Flow of general time independent non-Newtonian fluids in pipes

It is common practice to correlate non-Newtonian design data for flow in pipes of circular cross-section as plots of shear stress at the pipe wall R_w against flow characteristic $8u/d_i$. The data are plotted on either Cartesian or log-log coordinates. The slope of a log-log plot at any point is the flow behaviour index n' which can be written either as

$$n' = \frac{\text{d} \ln R_w}{\text{d} \ln (8u/d_i)} \qquad (3.1-1)$$

or as

$$n' = \frac{\text{d} \ln [\Delta P/(4L/d_i)]}{\text{d} \ln (8u/d_i)} \qquad (3.1-2)$$

Figure (3.1–1) is a plot of $\Delta P/(4L/d_i)$ against $8u/d_i$ for a typical time independent non-Newtonian fluid flow in a pipe. For the laminar flow region the plot gives a single line independent of pipe size[1] and for the turbulent flow region a separate line for each pipe size. It is shown in Figure (3.1–1) that the onset of turbulence in pipe flow is accompanied by a sharp increase in the shear stress at the pipe wall.

Equation (3.1–2) can be rewritten either as

$$\frac{\Delta P}{4L/d_i} = K'_p \left(\frac{8u}{d_i}\right)^{n'} \qquad (3.1-3)$$

39

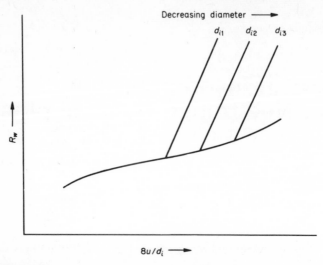

Figure (3.1–1)
Shear stress at the pipe wall against flow characteristic for a non-Newtonian fluid flowing in a pipe.

or as

$$\frac{\Delta P}{4L/d_i} = \left[K'_p \left(\frac{8u}{d_i} \right)^{n'-1} \right] \left(\frac{8u}{d_i} \right) \qquad (3.1\text{–}4)$$

where K'_p and n' are point values for a particular value of the flow characteristic $8u/d_i$.

By analogy with equation (2.4–7) for Newtonian fluids, the following equation can be written for non-Newtonian fluids.

$$\frac{\Delta P}{4L/d_i} = (\mu_a)_p \left(\frac{8u}{d_i} \right) \qquad (3.1\text{–}5)$$

where an apparent viscosity $(\mu_a)_p$ for pipe flow is defined as

$$\text{apparent viscosity } (\mu_a)_p = \frac{\text{shear stress at the pipe wall}}{\text{flow characteristic}}$$

or

$$(\mu_a)_p = \frac{R_w}{8u/d_i} \qquad (3.1\text{–}6)$$

By comparison of equation (3.1–5) with equation (3.1–4) a point value for the apparent viscosity $(\mu_a)_p$ can be written as

$$(\mu_a)_p = K'_p \left(\frac{8u}{d_i}\right)^{n'-1} \tag{3.1–7}$$

A Reynolds number for the flow of non-Newtonian fluids in pipes can be defined as

$$N_{RE} = \frac{\rho u d_i}{(\mu_a)_p} \tag{3.1–8}$$

Thus a point value for the Reynolds number can be written in terms of equation (3.1–7) as

$$N_{RE} = \frac{\rho u d_i}{K'_p (8u/d_i)^{n'-1}} \tag{3.1–9}$$

which can also be written in the form[6]

$$N_{RE} = \frac{\rho u^{2-n'} d_i^{n'}}{m} \tag{3.1–10}$$

where $m = K'_p 8^{n'-1}$.

Other apparent viscosities can be defined as

$$\text{apparent viscosity } (\mu_a)_{pw} = \frac{\text{shear stress at the pipe wall}}{\text{shear rate at the pipe wall}}$$

$$(\mu_a)_{pw} = \frac{R_w}{\dot{\gamma}_w} \tag{3.1–11}$$

and

$$\text{apparent viscosity } (\mu_a)_{pm} = \frac{\text{mean shear stress in the pipe}}{\text{mean shear rate in the pipe}}$$

$$(\mu_a)_{pm} = \frac{R_m}{\dot{\gamma}_m} \tag{3.1–12}$$

For Newtonian fluids the flow characteristic in a pipe $8u/d_i$ is a linear function of the shear stress at the pipe wall R_w given by equation (2.4–9)

$$\frac{8u}{d_i} = \frac{R_w}{\mu} \tag{2.4–9}$$

For non-Newtonian fluids the relationship between $8u/d_i$ and R_w is more complex and for general time independent fluids can be written as

$$\frac{8u}{d_i} = \phi(R_w) \tag{3.1–13}$$

3.2   Shear rate at a pipe wall for general time independent non-Newtonian fluids

For Newtonian fluids the shear rate at the pipe wall $\dot{\gamma}_w$ is equal to the flow characteristic $8u/d_i$.

For general time independent non-Newtonian fluids, the shear rate at the pipe wall $\dot{\gamma}_w$ is given by the equation

$$\dot{\gamma}_w = \left(\frac{8u}{d_i}\right)\left(\frac{3n' + 1}{4n'}\right) \tag{3.2–1}$$

Equation (3.2–1) can be proved as follows.

It has already been shown that the flow characteristic $8u/d_i$ is related to the velocity gradient at a radial point in a pipe $-\mathrm{d}v_x/\mathrm{d}r$ by equation (2.10–4).

$$\frac{8u}{d_i} = \frac{32}{d_i^3} \int_0^{d_i/2} r^2 \left(\frac{-\mathrm{d}v_x}{\mathrm{d}r}\right) \mathrm{d}r \tag{2.10–4}$$

For general time independent non-Newtonian fluids, $8u/d_i$ is also a function of the shear stress at the pipe wall, R_w, written as in equation (3.1–13).

$$\frac{8u}{d_i} = \phi(R_w) \tag{3.1–13}$$

The velocity gradient at a radial point in the pipe $-\mathrm{d}v_x/\mathrm{d}r$ is a function of the shear stress at a radial point in the pipe R_{rx} written as in equation (1.4–4).

$$\frac{-\mathrm{d}v_x}{\mathrm{d}r} = f(R_{rx}) \tag{1.4–4}$$

At the pipe wall equation (1.4–4) becomes

$$\dot{\gamma}_w = \left(\frac{-\mathrm{d}v_x}{\mathrm{d}r}\right)_w = f(R_w) \tag{3.2–2}$$

It has already been shown that R_{rx}, R_w and r are related by equation (2.3–4).

$$r = \frac{R_{rx} d_i}{2R_w} \tag{2.3–4}$$

Differentiate equation (2.3–4) to give

$$dr = \frac{dR_{rx} d_i}{2R_w} \tag{3.2–3}$$

Substitute equations (3.1–13), (1.4–4), (2.3–4) and (3.2–3) into equation (2.10–4) and rearrange to give

$$R_w^3 \phi(R_w) = 4 \int_0^{R_w} (R_{rx})^2 \, f(R_{rx}) \, dR_{rx} \tag{3.2–4}$$

Differentiate equation (3.2–4) with respect to R_w to give

$$3R_w^2 \phi(R_w) + R_w^3 \phi'(R_w) = 4R_w^2 \, f(R_w) \tag{3.2–5}$$

and rearrange to

$$4f(R_w) = 3\phi(R_w) + R_w \phi'(R_w) \tag{3.2–6}$$

Equation (3.2–6) can also be written in the expanded form

$$4f(R_w) = 3\phi(R_w) + \frac{R_w}{dR_w} dR_w \frac{\phi'(R_w)}{\phi(R_w)} \phi(R_w) \tag{3.2–7}$$

However

$$\frac{\phi'(R_w)}{\phi(R_w)} dR_w = d \ln \phi(R_w) \tag{3.2–8}$$

and

$$\frac{dR_w}{R_w} = d \ln R_w \tag{3.2–9}$$

Substitute equations (3.2–8) and (3.2–9) into equation (3.2–7) to give

$$4f(R_w) = 3\phi(R_w) + \frac{d \ln \phi(R_w)}{d \ln R_w} \phi(R_w) \tag{3.2–10}$$

Substitute equations (3.1–13) and (3.2–2) into equation (3.2–10) to give

$$4\dot{\gamma}_w = \left(\frac{8u}{d_i}\right)\left[3 + \frac{d \ln (8u/d_i)}{d \ln R_w}\right] \qquad (3.2–11)$$

Rewrite equation (3.1–1) as

$$\frac{1}{n'} = \frac{d \ln (8u/d_i)}{d \ln R_w} \qquad (3.2–12)$$

and substitute in equation (3.2–11) to give

$$4\dot{\gamma}_w = \left(\frac{8u}{d_i}\right)\left(3 + \frac{1}{n'}\right) \qquad (3.2–13)$$

which can be rewritten in the form

$$\dot{\gamma}_w = \left(\frac{8u}{d_i}\right)\left(\frac{3n' + 1}{4n'}\right) \qquad (3.2–1)$$

Equation (3.2–1) is the Metzner and Reed modification of the Rabinowitsch equation[6,8] and it is valid for any time independent non-Newtonian fluid.

3.3 Pressure drop in pipes for general time independent non-Newtonian fluids in laminar flow

For any time independent non-Newtonian fluid, a plot of $\Delta P/(4L/d_i)$ against $8u/d_i$ can be obtained using either a capillary tube viscometer or a small pilot size pipeline[7]. The pressure gradient $\Delta P/L$ for a particular fluid in laminar flow in a production size pipeline can be calculated from the appropriate plot of $\Delta P/(4L/d_i)$ against $8u/d_i$.

A point value for the Reynolds number N_{RE} at a particular flow characteristic $8u/d_i$ can be calculated from equation (3.1–9) knowing point values of K'_p and n'.

$$N_{RE} = \frac{u\rho d_i}{K'_p(8u/d_i)^{n'-1}} \qquad (3.1–9)$$

A point value of the basic friction factor j_f for laminar flow can be calculated from equation (2.4–5)[6].

$$j_f = \frac{8}{N_{RE}} \qquad (2.4–5)$$

The pressure drop can then be calculated from equation (2.4–3) in the same way as for Newtonian fluids.

$$\Delta P = 8j_f \left(\frac{L}{d_i}\right) \frac{\rho u^2}{2} \qquad (2.4–3)$$

3.4 Pressure drop in pipes for general time independent non-Newtonian fluids in turbulent flow

The theory of turbulent flow in pipes is far less developed than that for laminar flow. Consequently, turbulent flow pressure drops can be predicted with much less accuracy than laminar flow pressure drops.

For the turbulent flow of general time independent non-Newtonian fluids in smooth cylindrical pipes, Dodge and Metzner[2] have suggested the equation

$$f = \frac{a}{N_{RE}^b} \qquad (3.4–1)$$

to calculate the Fanning friction factor f where a and b are functions of the flow behaviour index n'. Values of a and b for each value of n' are listed in Table (3.4–1).

TABLE (3.4–1)

n'	a	b
0.2	0.0646	0.349
0.3	0.0685	0.325
0.4	0.0712	0.307
0.6	0.0740	0.281
0.8	0.0761	0.263
1.0	0.0779	0.250
1.4	0.0804	0.231
2.0	0.0826	0.213

For the turbulent flow of time independent non-Newtonian fluids in smooth cylindrical pipes, Dodge and Metzner[2] have also written the von Karman equation in the following general form

$$\frac{1}{f^{\frac{1}{2}}} = \frac{4.0}{(n')^{0.75}} \log \left\{ N_{RE} f^{[1-(n'/2)]} \right\} - \frac{0.4}{(n')^{1.2}} \qquad (3.4–2)$$

Equation (3.4–2) reduces to equation (2.4–14), the Newtonian form of the von Karman equation when $n' = 1$

$$\frac{1}{f^{\frac{1}{2}}} = 4.0 \log (N_{RE} f^{\frac{1}{2}}) - 0.4 \qquad (2.4\text{–}14)$$

The pressure drop can then be calculated from equation (2.4–4) in the same way as for Newtonian fluids.

$$\Delta P = 4f \left(\frac{L}{d_i}\right) \frac{\rho u^2}{2} \qquad (2.4\text{–}4)$$

Example (3.4–1)

A general time independent non-Newtonian liquid of density 961 kg/m^3 flows in steady state with a mean linear velocity of 1.523 m/s through a tube 3.048 m long with an inside diameter of 0.0762 m. For the above conditions, the point pipe consistency coefficient is $1.48 (\text{N/m}^2)\text{s}^{0.3}$ or $1.48 [\text{kg/(s}^2\text{ m)}] \text{ s}^{0.3}$ and the flow behaviour index is 0.3. Calculate the point values of the apparent viscosity and Reynolds number and also the pressure drop in N/m^2 in the tube.

Calculations:

$$\text{apparent viscosity } (\mu_a)_p = K'_p \left(\frac{8u}{d_i}\right)^{n'-1} \qquad (3.1\text{–}7)$$

$$\text{Reynolds number } N_{RE} = \frac{\rho u d_i}{(\mu_a)_p} \qquad (3.1\text{–}8)$$

$$\text{flow characteristic } \frac{8u}{d_i} = \frac{(8)(1.523 \text{ m/s})}{0.0762 \text{ m}} = 159.9 \text{ s}^{-1}$$

$$n' = 0.3$$

$$\left(\frac{8u}{d_i}\right)^{0.3} = 4.583 \text{ s}^{-0.3}$$

$$\left(\frac{8u}{d_i}\right)^{(0.3-1)} = 0.02866 \text{ s}^{-(0.3-1)}$$

$$(\mu_a)_p = \{1.48[\text{kg/(s}^2\text{ m)}]\text{s}^{0.3}\} [0.02866 \text{ s}^{-(0.3-1)}]$$

$$= 0.04242 \text{ kg/(s m)} = 0.04242 \text{ N s/m}^2$$

$$N_{RE} = \frac{(961 \text{ kg/m}^3)(1.523 \text{ m/s})(0.0762 \text{ m})}{0.04242 \text{ kg/(s m)}}$$

$$= 2629$$

Fanning friction factor

$$f = \frac{a}{N_{RE}^{b}} \qquad (3.4\text{--}1)$$

from Table (3.4–1) for $n' = 0.3$, $a = 0.0685$ and $b = 0.325$

$$f = \frac{0.0685}{2629^{0.325}} = \frac{0.0685}{12.92} = 0.005302$$

$$\frac{1}{f^{\frac{1}{2}}} = \frac{4.0}{(n')^{0.75}} \log \left[N_{RE} f^{(1-(n'/2))} \right] - \frac{0.4}{(n')^{1.2}} \qquad (3.4\text{--}2)$$

$$N_{RE} f^{(1-(n'/2))} = (2629)[0.005302^{(1-0.15)}]$$

$$= 2629 \times 0.01163$$

$$= 30.58$$

$$\log 30.58 = 1.4854$$

$$(n')^{0.75} = (0.3)^{0.75} = 0.4053$$

$$(n')^{1.2} = (0.3)^{1.2} = 0.2358$$

$$\frac{4.0}{(n')^{0.75}} = \frac{4.0}{0.4053} = 9.869$$

$$\frac{0.4}{(n')^{1.2}} = \frac{0.4}{0.2358} = 1.696$$

$$\frac{1}{f^{\frac{1}{2}}} = (9.869)(1.4854) - 1.696$$

$$= 14.659 - 1.696$$

$$= 12.963$$

$$f^{\frac{1}{2}} = 0.07714$$

$$f = (0.07714)^2 = 0.005951$$

pressure drop $\Delta P = 4f \left(\frac{L}{d_i} \right) \frac{\rho u^2}{2} \qquad (2.4\text{--}4)$

$$\frac{L}{d_i} = \frac{3.048 \text{ m}}{0.0762 \text{ m}} = 40$$

$$\frac{\rho u^2}{2} = \frac{(961 \text{ kg/m}^3)(1.523 \text{ m/s})^2}{2}$$

$$= 1115 \text{ kg/(s}^2 \text{ m)}$$

$$= 1115 \text{ N/m}^2$$

$$\Delta P = 4f \left(\frac{L}{d_i}\right)\frac{\rho u^2}{2}$$

$$= (4)(0.005951)(40)(1115 \text{ N/m}^2)$$

$$= 1062 \text{ N/m}^2$$

3.5   Pressure drop in pipes for Bingham plastics in laminar flow

For time independent fluids the velocity gradient $-\mathrm{d}v_x/\mathrm{d}r$ is related to the shear stress R_{rx} at a radial point in a pipe by equation (1.4–4)

$$\frac{-\mathrm{d}v_x}{\mathrm{d}r} = f(R_{rx}) \tag{1.4–4}$$

For Bingham plastics in steady laminar flow in pipes of circular cross-section, equation (1.4–4) can be written in the form

$$\frac{-\mathrm{d}v_x}{\mathrm{d}r} = \frac{R_{rx} - R_B}{\mu_p} \tag{3.5–1}$$

where R_B is the yield stress for the Bingham plastic and μ_p is the coefficient of rigidity.

It has already been shown that the flow characteristic $8u/d_i$ is related to $-\mathrm{d}v_x/\mathrm{d}r$ by equation (2.10–4)

$$\frac{8u}{d_i} = \frac{32}{d_i^3} \int_0^{d_i/2} r^2 \left(\frac{-\mathrm{d}v_x}{\mathrm{d}r}\right) \mathrm{d}r \tag{2.10–4}$$

Substitute equation (3.5–1) in equation (2.10–4) to give

$$\frac{8u}{d_i} = \frac{4}{R_w^3} \int_{R_B}^{R_w} (R_{rx})^2 \frac{(R_{rx} - R_B)}{\mu_p} \mathrm{d}R_{rx} \tag{3.5–2}$$

The integration limits in equation (3.5–2) are R_B to R_w and not 0 to R_w since no differential flow takes place when R_{rx} is below the yield stress R_B. Thus a Bingham plastic moves through a pipe as a solid plug surrounded by liquid.

Equation (3.5–2) can be integrated to give the well known Buckingham equation

$$\frac{8u}{d_i} = \frac{R_w}{\mu_p}\left[1 - \frac{4}{3}\left(\frac{R_B}{R_w}\right) + \frac{1}{3}\left(\frac{R_B}{R_w}\right)^4\right]$$ (3.5–3)

where R_w is the shear stress at a pipe wall given by equation (2.2–2)

$$R_w = \frac{\Delta P}{4L/d_i}$$ (2.2–2)

Govier[3] developed a method for solving equation (3.5–3) for the pressure drop ΔP. He defined a modified Reynolds number N'_{RE} and a dimensionless yield number N_Y for Bingham plastics as follows:

$$N'_{RE} = \frac{\rho u d_i}{\mu_p}$$ (3.5–4)

$$N_Y = \frac{R_B d_i}{u\mu_p}$$ (3.5–5)

where the product of N'_{RE} and N_Y is the dimensionless Hedstrom number N_{HE}

$$N_{HE} = \frac{R_B d_i^2}{\mu_p}$$ (3.5–6)

Equation (2.4–4) for the pressure drop in a pipe can be written in the form

$$f = \frac{\Delta P d_i}{2u^2 L\rho}$$ (3.5–7)

where f is the dimensionless Fanning friction factor.

Govier[3] combined equations (3.5–3), (2.2–2), (3.5–4), (3.5–5) and (3.5–7) to give the equation

$$\frac{1}{f N'_{RE}} = \frac{1}{16} - \frac{N_Y}{6f N'_{RE}} + \frac{N_Y^4}{3(f N'_{RE})^4}$$ (3.5–8)

The product $f N'_{RE}$ is a unique function of N_Y.[4] Govier[3] has given the corresponding values of $f N'_{RE}$ and N_Y which are listed in Table (3.5–1). Values of f at any value of N_Y may be obtained from

TABLE (3.5–1)

N_Y	$fN_{RE'}$
0	16.00
1	18.67
2	21.32
3	23.97
5	29.20
10	41.94
20	66.42
30	90.09
50	136.1
100	247.5
200	463.8
300	676.4
500	1096
1000	2133
2000	4186
3000	6226
5000	10 290
10 000	20 410

Table (3.5–1) by dividing the product $f N'_{RE}$ by N'_{RE}. The pressure drop can then be calculated from equation (2.4–4).

$$\Delta P = 4f \left(\frac{L}{d_i}\right)\frac{\rho u^2}{2} \tag{2.4–4}$$

3.6   Flow of power law fluids in pipes

Power law fluids are those in which the shear stress R is related to the shear rate $\dot{\gamma}$ by equation (1.4–5)

$$R = K(\dot{\gamma})^n \tag{1.4–5}$$

For the shear stress at a pipe wall R_w and the shear rate at a pipe wall $\dot{\gamma}_w$, equation (1.4–5) can be written

$$R_w = K(\dot{\gamma}_w)^n \tag{3.6–1}$$

Equation (3.1–3) gives the relationship between the pressure drop and the flow characteristic in a pipe for general time independent non-Newtonian fluids.

$$\frac{\Delta P}{4L/d_i} = K'_p \left(\frac{8u}{d_i}\right)^{n'} \tag{3.1–3}$$

For power law fluids the parameters K'_p and n' in equation (3.1–3) are no longer point values but remain constant over a range of flow characteristics. For these fluids equation (3.1–3) can be written as

$$\frac{\Delta P}{4L/d_i} = K_p \left(\frac{8u}{d_i}\right)^n \qquad (3.6–2)$$

where K_p is the consistency coefficient for pipe flow and n is the power law index.

Equation (3.2–1) gives the shear rate at a pipe wall for general time independent fluids.

$$\dot{\gamma}_w = \left(\frac{8u}{d_i}\right)\left(\frac{3n' + 1}{4n'}\right) \qquad (3.2–1)$$

For power law fluids where n' is constant, equation (3.2–1) can be written as

$$\dot{\gamma}_w = \left(\frac{8u}{d_i}\right)\left(\frac{3n + 1}{4n}\right) \qquad (3.6–3)$$

Combine equations (3.6–1), (3.6–2) and (3.6–3) to give the following relationship between the general consistency coefficient K and the consistency coefficient for pipe flow K_p:

$$K_p = K\left(\frac{3n + 1}{4n}\right)^n \qquad (3.6–4)$$

Equation (3.1–6) defines an apparent viscosity

$$(\mu_a)_p = \frac{R_w}{8u/d_i} \qquad (3.1–6)$$

which for power law fluids obeying equation (3.6–2) can be written as

$$(\mu_a)_p = K_p\left(\frac{8u}{d_i}\right)^{n-1} \qquad (3.6–5)$$

For Newtonian fluid flow in pipes, the onset of turbulent flow occurs at a Reynolds number N_{RE} of approximately 2100. Dodge and Metzner[2] showed that for pseudoplastic fluids obeying the power law, the end of the laminar flow region in pipes occurred at a Reynolds number which increased as the power law index n decreased.

A Reynolds number for the flow of non-Newtonian fluids in pipes can be defined as

$$N_{RE} = \frac{\rho u d_i}{(\mu_a)_p} \qquad (3.1\text{–}8)$$

For power law fluids where

$$(\mu_a)_p = K_p \left(\frac{8u}{d_i}\right)^{n-1} \qquad (3.6\text{–}5)$$

the Reynolds number can be written either as

$$N_{RE} = \frac{\rho u d_i}{K_p (8u/d_i)^{n-1}} \qquad (3.6\text{–}6)$$

or as

$$N_{RE} = \frac{\rho u^{2-n} d_i^n}{m} \qquad (3.6\text{–}7)$$

where $m = K_p 8^{n-1}$.

Consider the same power law fluid flowing in two different sized pipes with inside diameters of d_{i1} and d_{i2} respectively. Let the corresponding mean linear velocities in the pipes be u_1 and u_2 respectively. In order to have the same Reynolds number in the two pipes as defined by equation (3.6–7), the following equations must hold

$$\frac{u_1}{u_2} = \left(\frac{d_{i1}}{d_{i2}}\right)^{-n/(2-n)} \qquad (3.6\text{–}8)$$

and

$$\frac{u_1}{u_2} = \left(\frac{d_{i2}}{d_{i1}}\right)^{n/(2-n)} \qquad (3.6\text{–}9)$$

Example (3.6–1)

A power law liquid of density 961 kg/m³ flows in steady state with a mean linear velocity of 1.523 m/s through a tube 2.67 m long with an inside diameter of 0.0762 m. For a pipe consistency coefficient of 4.46 (N/m²) sⁿ or 4.46 [kg/s² m)] sⁿ, calculate the apparent viscosity in N s/m², the Reynolds number, and the pressure drop in N/m² and in kN/m² in the tube for power law indices $n = 0.3$, 0.7, 1.0 and 1.5 respectively.

Calculations:

$$\text{apparent viscosity } (\mu_a)_p = K_p \left(\frac{8u}{d_i}\right)^{n-1} \qquad (3.6\text{–}5)$$

$$\text{Reynolds number } N_{RE} = \frac{\rho u d_i}{(\mu_a)_p} \tag{3.1-8}$$

$$\text{for laminar flow, pressure drop } \Delta P = \left(\frac{4L}{d_i}\right)(\mu_a)_p\left(\frac{8u}{d_i}\right)$$

$$= \left(\frac{4L}{d_i}\right)K_p\left(\frac{8u}{d_i}\right)^n$$

$$\text{flow characteristic } \frac{8u}{d_i} = \frac{(8)(1.523 \text{ m/s})}{0.0762 \text{ m}} = 159.9 \text{ s}^{-1}$$

$$\frac{4L}{d_i} = \frac{(4)(2.67 \text{ m})}{0.0762 \text{ m}} = 140.2$$

$$(\mu_a)_p = \{4.46[\text{kg/(s}^2\text{ m)}] \text{ s}^n\}(159.9 \text{ s}^{-1})^{n-1}$$

$$= (4.46)(159.9^{n-1})\text{kg/(s m)}$$

$$N_{RE} = \frac{(961 \text{ kg/m}^3)(1.523 \text{ m/s})(0.0762 \text{ m})}{(4.46)(159.9^{n-1})\text{kg/(s m)}}$$

$$= \frac{25.01}{159.9^{n-1}}$$

$$\Delta P = (140.2)[4.46(\text{N/m}^2) \text{ s}^n](159.9 \text{ s}^{-1})^n$$

$$= (625.4)(159.9^n) \text{ N/m}^2$$

n	0.3	0.7	1.0	1.5
159.9^n	4.583	34.89	159.9	2022
$159.9^{n-1} = \dfrac{159.9^n}{159.9}$	0.02866	0.2182	1.0	12.645
$(\mu_a)_p$, N s/m² $= (4.46)(159.9^{n-1})$	0.1278	0.9732	4.46	56.40
N_{RE} $= 25.01/159.9^{n-1}$	872.6	114.6	25.01	1.978
ΔP, N/m² $= (713.6)(159.9^n)$	2 866	21 820	100 000	1 265 000
ΔP, kN/m²	2.87	21.8	100	1,270
$\dfrac{\Delta P \text{ (power law liquid)}}{\Delta P \text{ (Newtonian liquid)}}$	0.0287	0.218	1.0	12.70

3.7   Velocity distribution for a power law fluid in laminar flow in a pipe

It has already been shown that the shear stress R_{rx} at a radial point r in a pipe is given by equation (2.3–2)

$$R_{rx} = \left(\frac{\Delta P}{L}\right)\frac{r}{2} \qquad (2.3\text{–}2)$$

Equation (2.3–2) is true irrespective of the nature of the fluid in the pipe.

For a power law fluid in laminar flow, R_{rx} is related to the velocity gradient $-dv_x/dr$ by equation (1.4–7) rewritten in the form

$$\dot{\gamma}_r = \frac{-dv_x}{dr} = \left(\frac{R_{rx}}{K}\right)^{1/n} \qquad (3.7\text{–}1)$$

Combine equations (2.3–2) and (3.7–1) to give

$$\frac{dv_x}{dr} = -\left(\frac{\Delta P}{2LK}\right)^{1/n} r^{1/n} \qquad (3.7\text{–}2)$$

and integrate over the whole pipe to give

$$v_x = \left(\frac{\Delta P}{4KL/d_i}\right)^{1/n}\left(\frac{n}{n+1}\right)\left(\frac{d_i}{2}\right)\left[1 - \left(\frac{2r}{d_i}\right)^{(n+1)/n}\right] \qquad (3.7\text{–}3)$$

Equation (3.7–3) gives the point linear velocity v_x at any radial distance r in a pipe for steady laminar flow of a power law fluid through a pipe of circular cross-section.

For Newtonian fluids $n = 1$, $K = \mu$ and equation (3.7–2) becomes

$$v_x = \left(\frac{\Delta P}{L}\right)\frac{d_i^2}{16\mu}\left[1 - \left(\frac{2r}{d_i}\right)^2\right] \qquad (2.7\text{–}3)$$

At the pipe wall equation (3.7–1) can be written as

$$\dot{\gamma}_w = \left(\frac{R_w}{K}\right)^{1/n} \qquad (3.7\text{–}4)$$

where $\dot{\gamma}_w$ is the shear rate at the pipe wall which for a power law fluid is also given by equation (3.6–3)

$$\dot{\gamma}_w = \left(\frac{8u}{d_i}\right)\left(\frac{3n+1}{4n}\right) \qquad (3.6\text{–}3)$$

The corresponding shear stress at the pipe wall R_w is given by equation (2.2–2)

$$R_w = \frac{\Delta P}{4L/d_i} \qquad (2.2–2)$$

Combine equations (2.2–2), (3.6–3) and (3.7–4) to give

$$\left(\frac{8u}{d_i}\right)\left(\frac{3n+1}{4n}\right) = \left(\frac{\Delta P}{4KL/d_i}\right)^{1/n} \qquad (3.7–5)$$

Substitute equation (3.7–5) into equation (3.7–3) to give equation (3.7–6) which relates the linear velocity v_x at any radial distance r to the mean linear velocity u in a pipe of circular cross-section.

$$v_x = u\left(\frac{3n+1}{n+1}\right)\left[1 - \left(\frac{2r}{d_i}\right)^{(n+1)/n}\right] \qquad (3.7–6)$$

The effect of the variation of the power law index n on the shape of the laminar flow velocity profile in a pipe of circular cross-section is shown in Figure (3.7–1)[5]. For pseudoplastic fluids $n < 1$ and for dilatant fluids $n > 1$.

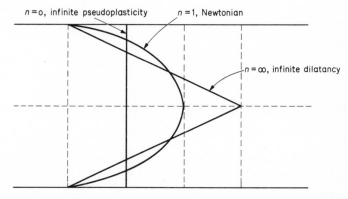

Figure (3.7–1)
Velocity profile for power law fluids.

3.8   Velocity distribution for a power law fluid in turbulent flow in a pipe

Dodge and Metzner[2] gave equations for the turbulent flow velocity profile of power law fluids in pipes. For the central flow region, the appropriate equation is

$$v^+ = \frac{5.66}{n^{0.75}} \log z^+ - \frac{0.40}{n^{1.2}} + \frac{2.458}{n^{0.75}}\left[1.960 + 1.255n \right.$$
$$\left. - 1.628n \log \left(3 + \frac{1}{n} \right) \right] \tag{3.8-1}$$

where v^+ and z^+ are a dimensionless velocity and distance respectively, defined as follows:

$$v^+ = \frac{v_x}{v^*}$$

where

$$v^* = \sqrt{\frac{R}{\rho}} \tag{2.9-4}$$

and

$$z^+ = \frac{\rho v^{*[1 - (n/2)]} z^n}{K} \tag{3.8-2}$$

For Newtonian fluids, $n = 1$ and $K = \mu$ and equation (3.8-2) reduces to

$$z^+ = \frac{v^* z}{\eta}$$

For Newtonian fluids, equation (3.8-1) reduces to

$$v^+ = 5.66 \log z^+ + 5.1 \tag{3.8-3}$$

which does not differ significantly from equation (2.9-16) which can also be rewritten as

$$v^+ = 5.75 \log z^+ + 5.5 \tag{3.8-4}$$

since $2.5 \ln z^+ = 5.75 \log z^+$.

For Newtonian fluids in the laminar sublayer adjacent to the wall

$$v^+ = z^+ \tag{2.9-6}$$

The corresponding equation for power law fluids is[2]

$$v^+ = (z^+)^{1/n} \tag{3.8-5}$$

3.9 Expansion and contraction losses for power law fluids

Wilkinson[9] showed that for power law fluids, the loss in pressure due to sudden expansion from a smaller to a larger diameter pipe of circular cross-section is

$$\Delta P_e = \rho\left(\frac{Q}{S_1}\right)^2\left(\frac{3n+1}{2n+1}\right)\left[\frac{n+3}{2(5n+3)}\left(\frac{S_1}{S_2}\right)^2 - \left(\frac{S_1}{S_2}\right) + \frac{3(2n+1)}{2(5n+3)}\right]$$

(3.9–1)

In equation (3.9–1), Q is the volumetric flow rate, ρ the density and n the power law index of the fluid. The smaller and the larger pipes have cross-sectional flow areas of S_1 and S_2 and inside diameters of d_{i1} and d_{i2} respectively.

$$\frac{S_1}{S_2} = \frac{d_{i1}^2}{d_{i2}^2}$$

(3.9–2)

When $n = 0$, equation (3.9–1) reduces to

$$\Delta P_e = \frac{\rho u_1^2}{2}\left[1 - \left(\frac{d_{i1}}{d_{i2}}\right)^2\right]^2$$

(2.5–2)

where u_1 is the mean linear velocity in the smaller diameter pipe. Equation (2.5–1) is identical with the approximate expression for the turbulent flow of a Newtonian fluid when the velocity profile is assumed to be flat.

Currently, there is no reliable expression for the pressure loss due to contraction for any kind of non-Newtonian fluid.

REFERENCES

(1) Alves, G. E., Boucher, D. F., and Pigford, R. L., Chem. Eng. Prog., 48, 385 (1952).
(2) Dodge, D. W., and Metzner, A. B., A.I.Ch.E.J., 5, 189 (1959).
(3) Govier, G. W., Chem. Eng., 66, No 17 (1959).
(4) Hedstrom, B. O. A., Ind. Eng. Chem., 44, 651 (1952).
(5) Holland, F. A., and Chapman, F. S., Pumping of Liquids, p. 68, New York, Reinhold Publishing Corporation, 1966.
(6) Metzner, A. B., and Reed, J. C., A.I.Ch.E.J., 1, 434 (1955).
(7) Potts, W. E., Brinkerhoff, R., Chapman, F. S., and Holland, F. A., Chem. Eng., 72, No 6 (1965).
(8) Rabinowitsch, B., Z. Phys. Chem., 145A, 1 (1929).
(9) Wilkinson, W. L., Non-Newtonian Fluids, p. 76, London, Pergamon Press, 1960.

4
Pumping of liquids

4.1 Pumps and pumping

Pumps are devices for supplying energy or head to a flowing liquid in order to overcome head losses due to friction and also, if necessary, to raise the liquid to a higher level.[2] The head imparted to a flowing liquid by a pump is known as the total head Δh. If a pump is placed between points 1 and 2 in a pipeline, the heads for steady flow are related by equation (1.6–7)

$$\left(z_2 + \frac{P_2}{\rho_2 g} + \frac{u_2^2}{2g\alpha_2} \right) - \left(z_1 + \frac{P_1}{\rho_1 g} + \frac{u_1^2}{2g\alpha_1} \right) = \Delta h - h_f \quad (1.6\text{–}7)$$

In equation (1.6–7), z, $P/(\rho g)$, and $u^2/(2g\alpha)$ are the static, pressure and velocity heads respectively and h_f is the head loss due to friction. The dimensionless velocity distribution factor α is $\frac{1}{2}$ for laminar flow and approximately 1 for turbulent flow.

For a liquid of density ρ flowing with a constant mean linear velocity u through a pipeline of circular cross-section and constant diameter between points 1 and 2 separated by a pump, equation (1.6–7) can be written as

$$\left(z_2 + \frac{P_2}{\rho g} + \frac{u^2}{2g\alpha} \right) - \left(z_1 + \frac{P_1}{\rho g} + \frac{u^2}{2g\alpha} \right) = \Delta h - h_f \quad (4.1\text{–}1)$$

For the most part, pumps can be classified into centrifugal and positive displacement pumps.

4.2   System heads

The important heads to consider in a pumping system are the suction, discharge, total and available net positive suction heads. The following definitions are given in reference to the typical

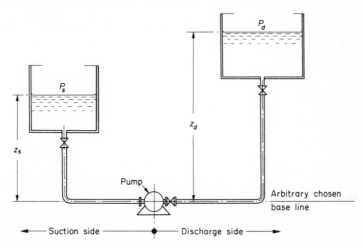

Figure (4.2–1)
Typical pumping system.

pumping system shown in Figure (4.2–1) where the arbitrarily chosen base line is the centreline of the pump.

Suction head:

$$h_s = z_s + \frac{P_s}{\rho g} - h_{fs} \qquad (4.2\text{–}1)$$

Discharge head:

$$h_d = z_d + \frac{P_d}{\rho g} + h_{fd} \qquad (4.2\text{–}2)$$

In equation (4.2–1), h_{fs} is the head loss due to friction, z_s is the static head and P_s is the gas pressure above the liquid in the tank on the suction side of the pump. If the liquid level on the suction side is below the centreline of the pump, z_s is negative.

In equation (4.2–2), h_{fd} is the head loss due to friction, z_d is the static head and P_d is the gas pressure above the liquid in the tank on the discharge side of the pump.

The total head Δh which the pump is required to impart to the flowing liquid is the difference between the discharge and suction heads.

Total head:

$$\Delta h = h_d - h_s \tag{4.2-3}$$

Equation (4.2–3) can be written in terms of equations (4.2–1) and (4.2–2) as

$$\Delta h = (z_d - z_s) + \frac{(P_d - P_s)}{\rho g} + (h_{fd} + h_{fs}) \tag{4.2-4}$$

The head losses due to friction are given by the equations

$$h_{fs} = 8j_f \left(\frac{\Sigma L_{es}}{d_i}\right) \frac{u^2}{2g} \tag{4.2-5}$$

and

$$h_{fd} = 8j_f \left(\frac{\Sigma L_{ed}}{d_i}\right) \frac{u^2}{2g} \tag{4.2-6}$$

where ΣL_{es} and ΣL_{ed} are the total equivalent lengths on the suction and discharge sides of the pump respectively.

The suction head h_s decreases and the discharge head h_d increases with increasing liquid flow rate because of the increasing value of the friction head loss terms h_{fs} and h_{fd}. Thus the total head Δh which the pump is required to impart to the flowing liquid increases with the liquid pumping rate.

Available net positive suction head:

$$NPSH = z_s + \frac{(P_s - P_{vp})}{\rho g} - h_{fs} \tag{4.2-7}$$

Equation (4.2–7) gives the head available to get the liquid through the suction piping. P_{vp} is the vapour pressure of the liquid being pumped at the particular temperature in question. The available net positive suction head $NPSH$ can also be written as

$$NPSH = h_s - \frac{P_{vp}}{\rho g} \tag{4.2-8}$$

The available $NPSH$ in a system should always be positive, i.e., the suction head must always be capable of overcoming the vapour pressure. Since the frictional head loss h_{fs} increases with increasing liquid flow rate, the available *NPSH* decreases with increasing liquid pumping rate. At the boiling temperature of the liquid, P_s and P_{vp} are equal and the available *NPSH* becomes $z_s - h_{fs}$. In this case no suction lift is possible since z_s must be positive. If the term $(P_s - P_{vp})$ is sufficiently large liquid can be lifted from below the centreline of the pump. In this case z_s is negative.

If the heads in equations (4.2–1), (4.2–2), (4.2–3), (4.2–4) and (4.2–7) are in m, then the pressures will be in N/m², the density in kg/m³ and the gravitational acceleration $g = 9.81$ m/s².

4.3   Centrifugal pumps

In centrifugal pumps, energy or head is imparted to a flowing liquid by centrifugal action. The most common type of centrifugal pump is the volute pump. In volute pumps, liquid enters near the axis of a high-speed impeller and is thrown radially outward into a progressively widening spiral casing as shown in Figure (4.3–1).

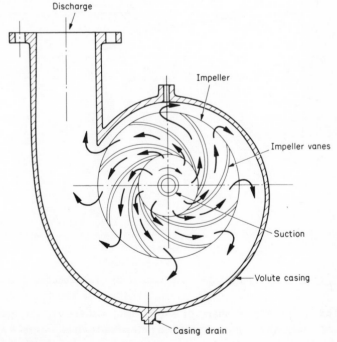

Figure (4.3–1)
Volute centrifugal pump casing design.

The impeller vanes are curved to ensure a smooth flow of liquid. The velocity head imparted to the liquid is gradually converted into pressure head as the velocity of the liquid is reduced. The efficiency of this conversion is a function of the design of the impeller and casing and the physical properties of the liquid.

The performance of a centrifugal pump for a particular rotational speed of the impeller and liquid viscosity is represented by plots of total head against capacity, power against capacity and required *NPSH* against capacity. These are known as characteristic curves of the pump. Characteristic curves have a variety of shapes depending on the geometry of the impeller and pump casing. Pump manufacturers normally supply these curves only for operation with water. However, methods are available for plotting curves for other viscosities from the water curves.[6]

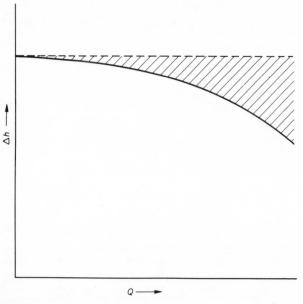

Figure (4.3–2)
Total head against capacity characteristic curve for a volute centrifugal pump.

The most common shape of a total head against capacity curve for a conventional volute centrifugal pump is shown in Figure (4.3–2) where Δh is the total head developed by the pump and Q is the volumetric flow rate of liquid or capacity. The maximum total head developed by the pump is at zero capacity. As the liquid throughput is increased, the total head developed decreases. The pump can operate at any point on the Δh against Q curve. Any individual Δh against Q curve is only true for a particular rotational speed of the impeller and liquid viscosity. As the liquid viscosity increases the Δh against Q curve becomes steeper. Thus the shaded area in Figure (4.3–2) increases as the liquid viscosity increases.

The total head Δh developed by a centrifugal pump at a particular capacity Q is independent of the liquid density. Thus the higher the density of the liquid, the higher the pressure ΔP developed by the pump. The relationship between ΔP and Δh is given by equation (4.3–1).

$$\Delta P = \rho \Delta h\, g \qquad (4.3\text{–}1)$$

Thus if a centrifugal pump develops a total head of 100 m when pumping a liquid of density $\rho = 1000\ \text{kg/m}^3$, the pressure developed is

$$\Delta P = (1000\ \text{kg/m}^3)(100\ \text{m})(9.81\ \text{m/s}^2)$$
$$= 981\ 000\ \text{N/m}^2$$

If the liquid has a density $\rho = 917\ \text{kg/m}^3$, the pressure developed when $\Delta h = 100$ m is

$$\Delta P = (917\ \text{kg/m}^3)(100\ \text{m})(9.81\ \text{m/s}^2)$$
$$= 900\ 000\ \text{N/m}^2$$

Equation (4.3–1) shows that when a centrifugal pump runs on air, the pressure developed is very small. In fact, a conventional centrifugal pump can never prime itself when operating on a suction lift.

In a particular system, a centrifugal pump can only operate at one point on the Δh against Q curve and that is the point where the pump Δh against Q curve intersects with the system Δh against Q curve as shown in Figure (4.3–3).

Equation (4.2–4) gives the system total head at a particular liquid flow rate.

$$\Delta h = (z_d - z_s) + \frac{(P_d - P_s)}{\rho g} + (h_{fd} + h_{fs}) \qquad (4.2\text{–}4)$$

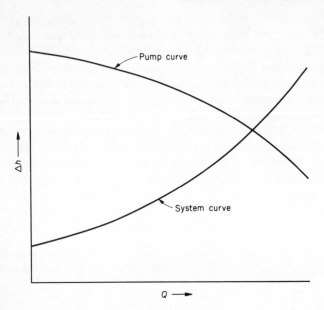

Figure (4.3–3)
System and pump total head against capacity curves.

Combine equation (4.2–4) with equations (4.2–5) and (4.2–6), which give the frictional head losses h_{f_s} and h_{f_d} respectively, to give

$$\Delta h = (z_d - z_s) + \frac{(P_d - P_s)}{\rho g} + 8j_f\left[\frac{(\Sigma L_{es} + \Sigma L_{ed})}{d_i}\right]\frac{u^2}{2g} \quad (4.3\text{–}2)$$

The mean linear velocity u of the liquid is related to the volumetric flow rate or capacity Q by equation (2.7–8)

$$u = \frac{Q}{\pi d_i^2/4} \quad (2.7\text{–}8)$$

Substitute equation (2.7–8) into equation (4.3–2) to give

$$\Delta h = (z_d - z_s) + \frac{(P_d - P_s)}{\rho g} + \frac{4j}{g}\left[\frac{(\Sigma L_{es} + \Sigma L_{ed})}{d_i}\right]\left(\frac{Q}{\pi d_i^2/4}\right)^2 \quad (4.3\text{–}3)$$

For laminar flow, the basic friction factor j_f is given by equation (2.4–5)

$$j_f = \frac{8}{N_{RE}} \quad (2.4\text{–}5)$$

For laminar flow substitute equation (2.4–5) into equation (4.3–2) to give

$$\Delta h = (z_d - z_s) + \frac{(P_d - P_s)}{\rho g} + \left(\frac{32\mu}{\rho d_i g}\right)\left[\frac{(\Sigma L_{es} + \Sigma L_{ed})}{d_i}\right]u \qquad (4.3\text{–}4)$$

Combine equations (2.7–8) and (4.3–4) and write

$$\Delta h = (z_d - z_s) + \frac{(P_d - P_s)}{\rho g} + \left(\frac{32\mu}{\rho d_i g}\right)\left[\frac{(\Sigma L_{es} + \Sigma L_{ed})}{d_i}\right]\left(\frac{Q}{\pi d_i^2/4}\right)$$
$$(4.3\text{–}5)$$

The system Δh against Q curve shown in Figure (4.3–3) can be plotted using equation (4.3–3) to calculate the values of the system total head Δh at each volumetric flow rate of liquid or capacity Q. Equation (4.3–5) shows that for laminar flow the total head Δh increases linearly with capacity Q. Thus for laminar flow, the system Δh against Q curve is a straight line.

In the above discussion it is assumed that the available $NPSH$ in the system is adequate to support the flow rate of liquid into the suction side of the pump. If the available $NPSH$ is less than that required by the pump, cavitation occurs and the normal curves do not apply. In cavitation, some of the liquid in the suction line vaporizes. As the vapour bubbles are carried into higher pressure regions of the pump they collapse, resulting in noise and vibration. High-speed pumps are more prone to cavitation than low-speed pumps.

Figure (4.3–4) shows a typical relationship between the available $NPSH$ in the system and the $NPSH$ required by the pump as the volumetric flow rate of liquid or capacity Q is varied. The $NPSH$ required by a centrifugal pump increases approximately with the square of the liquid throughput. The available $NPSH$ in a system can be calculated from equation (4.2–7) rewritten in terms of equations (2.7–8) and (4.2–5) as

$$NPSH = z_s + \frac{(P_s - P_{vp})}{\rho g} - \frac{4j}{g}\left(\frac{\Sigma L_{es}}{d_i}\right)\left(\frac{Q}{\pi d_i^2/4}\right)^2 \qquad (4.2\text{–}9)$$

Equation (4.2–9) shows that the available $NPSH$ in a system decreases as the liquid throughput increases because of the greater frictional head losses.

A centrifugal pump will operate normally at a point on its total head against capacity characteristic curve until the available

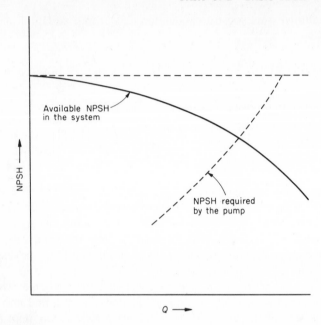

Figure (4.3–4)
Available and required net positive suction heads against capacity in a pumping system.

NPSH falls below the required *NPSH* curve. Beyond this point, the total head generated by a centrifugal pump falls drastically as shown in Figure (4.3–5) as the pump begins to operate in cavitation conditions.

In centrifugal pump systems, a throttling valve is located on the discharge side of the pump. When this valve is throttled, the system Δh against Q curve is altered to incorporate the increased frictional head loss. The effect of throttling is illustrated in Figure (4.3–6). Throttling can be used to decrease cavitation.

System total heads should be estimated as accurately as possible. Safety factors should never be added to these estimated total head values. This is illustrated by Figure (4.3–7). Suppose that OA_1 is the correct curve and that the centrifugal pump is required to operate at point A_1. Let a safety factor be added to the total head values to give a system curve OA_2. On the basis of curve OA_2, the manufacturer will supply a pump to operate at point A_2. However, since the true system curve is OA_1, the pump will operate at point A_3.

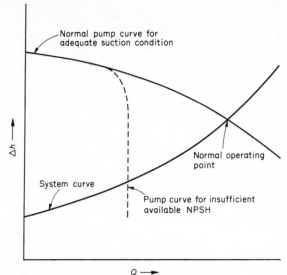

Figure (4.3–5)
Effect of insufficient available $NPSH$ on the performance of a centrifugal pump.

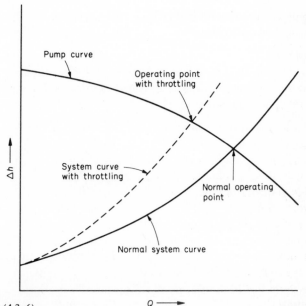

Figure (4.3–6)
Effect of throttling the discharge valve on the operating point of a centrifugal pump.

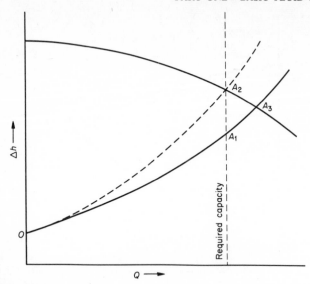

Figure (4.3–7)
Effect of adding a safety factor to the system total head against capacity curve

Not only is the capacity higher than that specified, but the pump motor may be overloaded.

Example (4.3–1)

Calculate the data for a system total head against capacity curve for the initial conditions of the system shown in Figure (4.2–1) given the following data:

dynamic viscosity of liquid	$\mu = 0.04\ \text{N s/m}^2$
density of liquid	$\rho = 1200\ \text{kg/m}^3$
static head on suction side of pump	$z_s = 3\ \text{m}$
static head on discharge side of pump	$z_d = 7\ \text{m}$
inside diameter of pipe	$d_i = 0.0526\ \text{m}$
pipe roughness	$\varepsilon = 0.000045\ \text{m}$
gas pressure above the liquid in the tank on the suction side of the pump	$P_s = \text{atmospheric pressure}$
gas pressure above the liquid in the tank on the discharge side of the pump	$P_d = \text{atmospheric pressure}$
total equivalent length on the suction side of the pump	$\Sigma L_{es} = 4.9\ \text{m}$

total equivalent length on the
    discharge side of the pump $\Sigma L_{ed} = 63.2$ m

In this case neglect entrance and exist losses although normally these should be included.

Calculations:

$$\text{Reynolds number } N_{RE} = \frac{\rho u d_i}{\mu} \qquad (2.1\text{--}1)$$

$$\rho = 1200 \text{ kg/m}^3$$

initially take $u = 1.0$ m/s

$d_i = 0.0526$ m

$\mu = 0.04 \text{ N s/m}^2 = 0.04 \text{ kg/(s m)}$

$$N_{RE} = \frac{(1200 \text{ kg/m}^3)(1.0 \text{ m/s})(0.0526 \text{ m})}{0.04 \text{ kg/(s m)}} = 1578$$

pipe roughness $\varepsilon = 0.000045$ m

$d_i = 0.0526$ m

$$\text{roughness factor } \frac{\varepsilon}{d_i} = \frac{0.000045 \text{ m}}{0.0526 \text{ m}} = 0.000856$$

from j_f against N_{RE} graph in Figure (2.4–1),

$$j_f = 0.00506 \quad \text{for} \quad N_{RE} = 1578 \quad \text{and} \quad \frac{\varepsilon}{d_i} = 0.000856$$

$$\frac{(\Sigma L_{es} + \Sigma L_{ed})}{d_i} = \frac{4.9 \text{ m} + 63.2 \text{ m}}{0.0526 \text{ m}} = 1294.7$$

$$\frac{u^2}{2g} = \frac{(1.0 \text{ m/s})^2}{(2)(9.81 \text{ m/s}^2)} = 0.05097 \text{ m}$$

$$\text{total head } \Delta h = (z_d - z_s) + \frac{(P_d - P_s)}{\rho g} + 8 j_f \left[\frac{(\Sigma L_{es} + \Sigma L_{ed})}{d_i} \right] \frac{u^2}{2g}$$
$$(4.3\text{--}2)$$

$$= 4 \text{ m} + 8(0.00506)(1294.7)(0.05097 \text{ m})$$

$$= 6.671 \text{ m}$$

$$\text{mean linear velocity } u = \frac{Q}{\pi d_i^2 / 4} \qquad (2.7\text{--}8)$$

$$\frac{\pi d_i^2}{4} = \frac{(3.142)(0.0526 \text{ m})^2}{4} = 0.002173 \text{ m}^2$$

$$\text{capacity } Q = u \frac{\pi d_i^2}{4}$$

$$= (1.0 \text{ m/s})(0.002173 \text{ m}^2)$$

$$= 0.002173 \text{ m}^3/\text{s} = 0.00217 \text{ m}^3/\text{s}$$

Repeat the calculations for other values of u and list as follows:

u m/s	N_{RE}	j_f	$u^2/2g$ m	$8j_f\left[\dfrac{(\Sigma L_{es} + \Sigma L_{ed})}{d_i}\right]\dfrac{u^2}{2g}$ m	Δh m	Q m³/s
0.5	789	0.01014	0.01274	1.338	5.3	0.00109
1.0	1 578	0.00506	0.05097	2.671	6.7	0.00217
1.5	2 367	—	0.1149	—	—	0.00326
2.0	3 156	0.0052	0.2039	10.98	15.0	0.00435
2.5	3 945	0.0050	0.3186	16.50	20.5	0.00544
3.0	4 734	0.0048	0.4487	22.31	26.3	0.00653

4.4 Centrifugal pump relations

The power P_E required in an ideal centrifugal pump can be expected to be a function of the liquid density ρ, the impeller diameter D and the rotational speed of the impeller N. If the relationship is assumed to be given by the equation

$$P_E = C\rho^a N^b D^c \tag{4.4–1}$$

then it can be shown by dimensional analysis[3] that

$$P_E = C_1 \rho N^3 D^5 \tag{4.4–2}$$

where C_1 is a constant which depends on the geometry of the system.

The power P_E is also proportional to the product of the volumetric flow rate Q and the total head Δh developed by the pump.

$$P_E = C_2 Q \Delta h \tag{4.4–3}$$

where C_2 is a constant.

The volumetric flow rate Q and the total head Δh developed by the pump are related to the rotational speed of the impeller N and the impeller diameter D by equations (4.4–4) and (4.4–5) respectively:

$$Q = C_3 N D^3 \qquad (4.4–4)$$

$$\Delta h = C_4 N^2 D^2 \qquad (4.4–5)$$

where C_3 and C_4 are constants.

Rewrite equation (4.4–5) in the form

$$\Delta h^{\frac{3}{2}} = C_4^{\frac{3}{2}} N^3 D^3 \qquad (4.4–6)$$

Combine equations (4.4–4) and (4.4–6) to give

$$\frac{N^2 Q}{\Delta h^{\frac{3}{2}}} = \text{a constant} \qquad (4.4–7)$$

or

$$\frac{N\sqrt{Q}}{\Delta h^{\frac{3}{4}}} = \text{a constant} \qquad (4.4–8)$$

When the rotational speed of the impeller N is in rpm, the volumetric flow rate Q is in U.S. gpm and the total head Δh developed by the pump is in ft, the constant in equation (4.4–8) is known as the specific speed N_s. The specific speed is used as an index of pump types and is always evaluated at the best efficiency point (bep) of the pump. Specific speed has the dimensions $(L/T^2)^{\frac{3}{4}}$.

Two different size pumps are said to be geometrically similar when the ratios of corresponding dimensions in one pump are equal to those of the other pump.[4] Geometrically similar pumps are said to be homologous. A set of equations known as the affinity laws govern the performance of homologous centrifugal pumps at various impeller speeds.

Consider a centrifugal pump with an impeller diameter D_1 operating at a rotational speed N_1 and developing a total head Δh_1. Consider an homologous pump with an impeller diameter D_2 operating at a rotational speed N_2 and developing a total head Δh_2.

Equations (4.4–4) and (4.4–5) for this case can be rewritten respectively in the form

$$\frac{Q_1}{Q_2} = \left(\frac{N_1}{N_2}\right)\left(\frac{D_1}{D_2}\right)^3 \qquad (4.4–9)$$

and

$$\frac{\Delta h_1}{\Delta h_2} = \left(\frac{N_1}{N_2}\right)^2\left(\frac{D_1}{D_2}\right)^2 \qquad (4.4–10)$$

Similarly equation (4.4–2) can be rewritten in the form

$$\frac{P_{E1}}{P_{E2}} = \left(\frac{N_1}{N_2}\right)^3 \left(\frac{D_1}{D_2}\right)^5 \qquad (4.4\text{–}11)$$

and by analogy with equation (4.4–10) the net positive suction heads for the two homologous pumps can be related by the equation

$$\frac{NPSH_1}{NPSH_2} = \left(\frac{N_1}{N_2}\right)^2 \left(\frac{D_1}{D_2}\right)^2 \qquad (4.4\text{–}12)$$

Equations (4.4–9), (4.4–10), (4.4–11) and (4.4–12) are the affinity laws for homologous centrifugal pumps.

For a particular pump where the impeller of diameter D_1 is replaced by an impeller with a slightly different diameter D_2, the following equations hold:[5]

$$\frac{Q_1}{Q_2} = \left(\frac{N_1}{N_2}\right)\left(\frac{D_1}{D_2}\right) \qquad (4.4\text{–}13)$$

$$\frac{\Delta h_1}{\Delta h_2} = \left(\frac{N_1}{N_2}\right)^2 \left(\frac{D_1}{D_2}\right)^2 \qquad (4.4\text{–}14)$$

and

$$\frac{P_{E1}}{P_{E2}} = \left(\frac{N_1}{N_2}\right)^3 \left(\frac{D_1}{D_2}\right)^3 \qquad (4.4\text{–}15)$$

If the characteristic performance curves are available for a centrifugal pump operating at a given rotation speed, equations (4.4–13), (4.4–14) and (4.4–15) enable the characteristic performance curves to be plotted for other operating speeds and for other slightly different impeller diameters.

Example (4.4–1)

A volute centrifugal pump with an impeller diameter of 0.02 m has the following performance data when pumping water at the best efficiency point:

$$\text{impeller speed } N = 58.3 \text{ rev/s}$$

$$\text{capacity } Q = 0.012 \text{ m}^3/\text{s}$$

$$\text{total head } \Delta h = 70 \text{ m}$$

$$\text{required net positive suction head } NPSH = 18 \text{ m}$$

$$\text{power } P_B = 12\,000 \text{ W}$$

Evaluate the performance data of an homologous pump with twice the impeller diameter operating at half the impeller speed.

Calculations:

Let subscripts 1 and 2 refer to the first and second pumps respectively.

ratio of impeller speeds $\dfrac{N_1}{N_2} = 2$

ratio of impeller diameters $\dfrac{D_1}{D_2} = \dfrac{1}{2}$

ratio of capacities

$$\frac{Q_1}{Q_2} = \left(\frac{N_1}{N_2}\right)\left(\frac{D_1}{D_2}\right)^3 \qquad (4.4\text{–}9)$$

$$= (2)(\tfrac{1}{8}) = \tfrac{1}{4}$$

capacity of second pump

$$Q_2 = 4Q_1 = (4)(0.012 \,\text{m}^3/\text{s})$$

$$= 0.048 \,\text{m}^3/\text{s}$$

ratio of total heads

$$\frac{\Delta h_1}{\Delta h_2} = \left(\frac{N_1}{N_2}\right)^2\left(\frac{D_1}{D_2}\right)^2 \qquad (4.4\text{–}10)$$

$$= (4)(\tfrac{1}{4}) = 1$$

total head of second pump

$$\Delta h_2 = \Delta h_1 = 70 \,\text{m}$$

ratio of powers

$$\frac{P_{E1}}{P_{E2}} = \left(\frac{N_1}{N_2}\right)^3\left(\frac{D_1}{D_2}\right)^5 \qquad (4.4\text{–}11)$$

$$= (8)(\tfrac{1}{32}) = \tfrac{1}{4}$$

Assume

$$\frac{P_{B1}}{P_{B2}} = \frac{P_{E1}}{P_{E2}}$$

$$\frac{P_{B1}}{P_{B2}} = \frac{1}{4}$$

power for second pump

$$P_{B2} = 4P_{B1} = (4)(12\,000\,\text{W})$$
$$= 48\,000\,\text{W}$$

ratio of required net positive suction heads

$$\frac{NPSH_1}{NPSH_2} = \left(\frac{N_1}{N_2}\right)^2\left(\frac{D_1}{D_2}\right)^2 \qquad (4.4\text{--}12)$$

$$= (4)(\tfrac{1}{4}) = 1$$

net positive suction head of second pump

$$NPSH_2 = NPSH_1 = 18\,\text{m}$$

4.5   Centrifugal pumps in series and in parallel

Diskind[1] determined the operating characteristics for centrifugal pumps in parallel and in series using a simple graphical method.

Consider two centrifugal pumps in parallel as shown in Figure (4.5–1). The total head for the pump combination Δh_T is the same as the total head for each pump, i.e.

$$\Delta h_T = \Delta h_1 = \Delta h_2 \qquad (4.5\text{--}1)$$

The volumetric flow rate or capacity for the pump combination Q_T is the sum of the capacities for the two pumps, i.e.

$$Q_T = Q_1 + Q_2 \qquad (4.5\text{--}2)$$

The operating characteristics for two pumps in parallel are obtained as follows:

(1) Draw the Δh against Q characteristic curves for each pump together with the system Δh_s against Q_s curve on the same plot as shown in Figure (4.5–1).
(2) Draw a horizontal constant total head line in Figure (4.5–1) which intersects the two pump curves at capacities Q_1 and Q_2 respectively, and the system curve at capacity Q_s.
(3) Add the values of Q_1 and Q_2 obtained in step (2) to give

$$Q_T = Q_1 + Q_2 \qquad (4.5\text{--}2)$$

(4) Compare Q_T from step (3) with Q_s from step (2). If they are not equal repeat steps (2), (3) and (4) until $Q_T = Q_s$. This is the operating point of the two pumps in parallel.

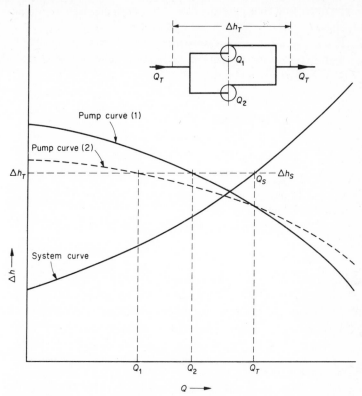

Figure (4.5–1)
Operating point for centrifugal pumps in parallel.

An alternative to this trial and error procedure for two pumps in parallel is to calculate Q_T from equation (4.5–2) for various values of the total head from known values of Q_1 and Q_2 at these total heads. The operating point for stable operation is at the intersection of the Δh_T against Q_T curve with the Δh_s against Q_s curve.

Consider two centrifugal pumps in series as shown in Figure (4.5–2). The total head for the pump combination Δh_T is the sum of the total heads for the two pumps, i.e.

$$\Delta h_T = \Delta h_1 + \Delta h_2 \qquad (4.5–3)$$

The volumetric flow rate or capacity for the pump combination Q_T is the same as the capacity for each pump, i.e.

$$Q_T = Q_1 = Q_2 \qquad (4.5–4)$$

The operating characteristics for two pumps in series are obtained as follows:

(1) Draw the Δh against Q characteristic curves for each pump together with the system Δh_s against Q_s curve on the same plot as shown in Figure (4.5–2).

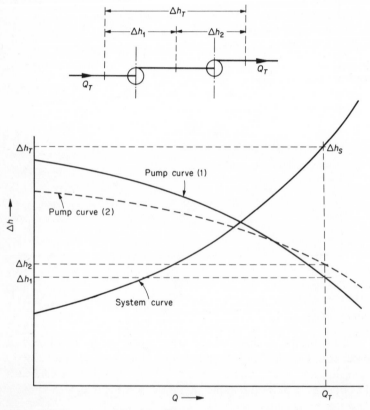

Figure (4.5–2)
Operating point for centrifugal pumps in series.

(2) Draw a vertical constant capacity line in Figure (4.5–2) which intersects the two pump curves at total heads Δh_1 and Δh_2 respectively, and the system curve at total head Δh_s.

(3) Add the values of Δh_1 and Δh_2 obtained in step (2) to give

$$\Delta h_T = \Delta h_1 + \Delta h_2 \qquad (4.5–3)$$

(4) Compare Δh_T from step (3) with Δh_s from step (2). If they are not equal repeat steps (2), (3) and (4) until $\Delta h_T = \Delta h_s$. This is the operating point of the two pumps in series.

An alternative to this trial and error procedure for two pumps in series is to calculate Δh_T from equation (4.5–3) for various values of the capacity from known values of Δh_1 and Δh_2 at these capacities. The operating point for stable operation is at the intersection of the Δh_T against Q_T curve with the Δh_s against Q_s curve.

The piping and valves may be arranged to enable two centrifugal pumps to be operated either in series or in parallel. For two identical pumps, series operation gives a total head of $2\Delta h$ at a capacity Q and parallel operation gives a capacity of $2Q$ at a total head Δh. The efficiency of either the series or parallel combination is practically the same as for a single pump.

4.6 Positive displacement pumps

For the most part, positive displacement pumps can be classified either as rotary pumps or as reciprocating pumps. However, pumps do exist which exhibit some of the characteristics of both types.

Rotary pumps forcibly transfer liquid through the action of rotating gears, lobes, vanes, screws etc., which operate inside a rigid container. Normally, pumping rates are varied by changing the rotational speed of the rotor. Rotary pumps do not require valves in order to operate.

Reciprocating pumps forcibly transfer liquid by changing the internal volume of the pump. Pumping rates are varied by altering either the frequency or the length of the stroke. Valves are required on both the suction and discharge sides of the pump.

One of the most common rotary pumps is the external gear pump illustrated in Figure (4.6–1). The fixed casing contains two meshing gears of equal size. The driving gear is coupled to the drive shaft which transmits the power from the motor. The idler gear runs free. As the rotating gears unmesh they create a partial vacuum which causes liquid from the suction line to flow into the pump. Liquid is carried through the pump between the rotating gear teeth and the fixed casing. The meshing of the rotating gears generates an increase in pressure which forces the liquid into the outlet line. In principle an external gear pump can discharge liquid either way depending on the direction of the gear rotation. In practice, external gear pumps are equipped with relief valves to limit the discharge pressures

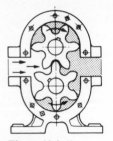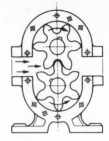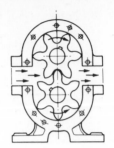

Figure (4.6–1)
Operation of an external gear pump.

generated since they cannot be operated against a closed discharge
without damage to the pump. In this case the direction of gear
rotation is fixed and is clearly marked on the pump.

External gear pumps are self-priming since the rotating gears are
capable of pumping air. They give a constant delivery of liquid for a
set rotor speed with negligible pulsations. Changes in capacity are
small with variations in discharge pressure and liquid viscosity.
External gear pumps depend on the liquid pumped to lubricate
the internal moving parts. They can be damaged if run dry. In order
to provide for changes in pumping rate, variable speed drives are
required.

Since close clearances are essential between the moving parts,
alignment is critical. Some leakage occurs between the discharge
and suction sides of a pump through the clearances. This is known as
slip. Slip increases with pressure difference across the pump and
decreases with increasing liquid viscosity. Since slip is independent
of pump speed, it is an advantage to pump low viscosity liquids at
high speeds. Slip is negligible for high viscosity liquids and in fact
external gear pumps are often used as metering pumps.

Rotary pumps can normally be divided into two classes: small
liquid cavity high-speed pumps and large liquid cavity low-speed
pumps.

4.7   Pumping efficiencies

The liquid power P_E can be defined as the rate of useful work done
on the liquid. It is given by the equation

$$P_E = Q \, \Delta P \qquad\qquad (4.7–1)$$

If the volumetric flow rate Q is in m^3/s and the pressure developed by the pump ΔP is in N/m^2, the liquid power P_E is in N m/s or W. The pressure developed by the pump ΔP is related to the total head developed by the pump Δh by equation (4.3–1).

$$\Delta P = \rho \, \Delta h \, g \qquad\qquad (4.3–1)$$

Substitute equation (4.3–1) into equation (4.7–1) to give

$$P_E = \rho Q \, \Delta h \, g \qquad\qquad (4.7–2)$$

which can also be written as

$$P_E = M \, \Delta h \, g \qquad\qquad (4.7–3)$$

since $M = \rho Q$ the flow rate in mass per unit time. If M is in kg/s, Δh is in m and the gravitational acceleration $g = 9.81$ m/s^2, P_E is in W.

The brake power P_B can be defined as the actual power delivered to the pump by the prime mover. It is the sum of liquid power and friction power and is given by the equation

$$P_B = P_E\left(\frac{100}{E}\right) \qquad\qquad (4.7–4)$$

where E is the mechanical efficiency expressed in per cent.

The mechanical efficiency decreases as the liquid viscosity and hence the frictional losses increase. The mechanical efficiency is also decreased by power losses in gears, bearings, seals etc. In rotary pumps contact between the rotor and the fixed casing increases power losses and decreases the mechanical efficiency. These losses are not proportional to pump size. Relatively large pumps tend to have the best efficiencies whilst small pumps usually have low efficiencies. Furthermore high-speed pumps tend to be more efficient than low-speed pumps. In general, high efficiency pumps have high NPSH requirements. Sometimes a compromise may have to be made between efficiency and NPSH.

Another efficiency which is important for positive displacement pumps is the volumetric efficiency. This is the delivered capacity per cycle as a percentage of the true displacement per cycle. If no slip occurs, the volumetric efficiency of the pump is 100 per cent. For zero pressure difference across the pump, there is no slip and the delivered capacity is the true displacement. The volumetric efficiency of a pump is reduced by the presence of entrained air or gas in the pumped liquid. It is important to know the volumetric efficiency of a positive displacement pump when it is to be used for metering.

4.8   Factors in pump selection

The selection of a pump depends on many factors which include the required rate and properties of the pumped liquid and the desired location of the pump.

In general, high viscosity liquids are pumped with positive displacement pumps. Centrifugal pumps are not only very inefficient when pumping high viscosity liquids but their performance is very sensitive to changes in liquid viscosity. A high viscosity also leads to high frictional head losses and hence a reduced available *NPSH*. Since the latter must always be greater than the *NPSH* required by the pump, a low available *NPSH* imposes a severe limitation on the choice of a pump. Liquids with a high vapour pressure also reduce the available *NPSH*. If these liquids are pumped at a high temperature, this may cause the gears to seize in a close clearance gear pump.

If the pumped liquid is pseudoplastic, its apparent viscosity will decrease with an increase in shear rate and hence pumping rate. It is therefore an advantage to use high-speed pumps to pump pseudoplastic liquids and in fact centrifugal pumps are frequently used. In contrast, the apparent viscosity of a dilatant liquid will increase with an increase in shear rate and hence pumping rate. It is therefore an advantage to use large cavity positive displacement pumps with a low cycle speed to pump dilatant liquids.

Some liquids can be permanently damaged by subjecting them to high shear in a high-speed pump. For example, certain liquid detergents can be broken down into two phases if subject to too much shear. Even though these detergents may exhibit pseudoplastic characteristics they should be pumped with relatively low-speed pumps.

Wear is a more serious problem with positive displacement pumps than with centrifugal pumps. Liquids with poor lubricating qualities increase the wear on a pump. Wear is also caused by corrosion and by the pumping of liquids containing suspended solids which are abrasive.

In general, centrifugal pumps are less expensive, last longer and are more robust than positive displacement pumps. However, they are unsuitable for pumping high viscosity liquids and when changes in viscosity occur.

REFERENCES

(1) Diskind, T., Chem. Eng., 66, No 22 (1959).
(2) Holland, F. A., and Chapman, F. S., *Pumping of Liquids*, p. 121, New York, Reinhold Publishing Corporation, 1966.
(3) Ibid., p. 193.
(4) Ibid., p. 194.
(5) Ibid., p. 196.
(6) Ibid., p. 249.

5
Mixing of liquids in tanks

5.1 Mixers and mixing

Quillen[19] defined mixing as the 'intermingling of two or more dissimilar portions of a material, resulting in the attainment of a desired level of uniformity, either physical or chemical, in the final product'. Since natural diffusion in liquids is relatively slow, liquid mixing is most commonly accomplished by rotating an agitator in the liquid confined in a tank. It is possible to waste much of this input of mechanical energy if the wrong kind of agitator is used. Parker[18] defined agitation as 'the creation of a state of activity such as flow or turbulence, apart from any mixing accomplished'.

A rotating agitator generates high velocity streams of liquid which in turn entrain stagnant or slower moving regions of liquid resulting in uniform mixing by momentum transfer. As the viscosity of the liquid is increased, the mixing process becomes more difficult since frictional drag retards the high velocity streams and confines them to the immediate vicinity of the rotating agitator.

In general, agitators can be classified into the following two groups:
(1) Agitators with a small blade area which rotate at high speeds. These include turbines and marine type propellers.
(2) Agitators with a large blade area which rotate at low speeds. These include anchors, paddles and helical screws.

The second group is more effective than the first in the mixing of high viscosity liquids.

The mean shear rate produced by an agitator in a mixing tank $\dot{\gamma}_m$ is proportional to the rotational speed of the agitator N.[15]

Thus

$$\dot{\gamma}_m = kN \qquad (5.1\text{-}1)$$

where k is a dimensionless proportionality constant for a particular system.

For a liquid mixed in a tank with a rotating agitator, the shear rate is greatest in the immediate vicinity of the agitator. In fact the shear rate decreases exponentially with distance from the agitator.[17] Thus the shear stresses and strains vary greatly throughout an agitated liquid in a tank. Since the dynamic viscosity of a Newtonian liquid is independent of shear at a given temperature, its viscosity will be the same at all points in the tank. In contrast the apparent viscosity of a non-Newtonian liquid varies throughout the tank. This in turn significantly influences the mixing process. For pseudoplastic liquids, the apparent viscosity is at a minimum in the immediate vicinity of the agitator. The progressive increase in the apparent viscosity of a pseudoplastic liquid with distance away from the agitator tends to dampen eddy currents in the mixing tank. In contrast, for dilatant liquids, the apparent viscosity is at a maximum in the immediate vicinity of the agitator. In general pseudoplastic and dilatant liquids should be mixed using high and low speed agitators respectively.

It is desirable to produce a particular mixing result in the minimum time t and with the minimum input of power per unit volume P_A/V. Thus an efficiency function E can be defined as

$$E = \left(\frac{1}{P_A/V}\right)\left(\frac{1}{t}\right) \qquad (5.1\text{-}2)$$

5.2 Small blade high speed agitators

Small blade high speed agitators are used to mix low to medium viscosity liquids. Two of the most common types are the 6-blade flat blade turbine and the marine type propeller shown in Figures (5.2–1) and (5.2–2) respectively. Flat blade turbines used to mix liquids in baffled tanks produce radial flow patterns primarily perpendicular to the vessel wall as shown in Figure (5.2–3). In contrast marine type propellers used to mix liquids in baffled tanks produce axial flow patterns primarily parallel to the vessel wall as shown in Figure (5.2–4). Marine type propellers and flat blade

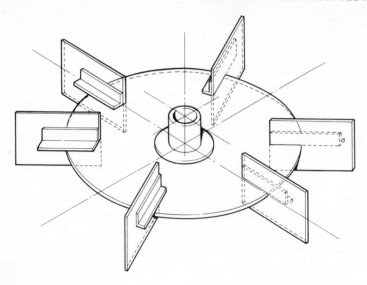

Figure (5.2–1)
Six blade flat blade turbine

turbines are suitable to mix liquids with dynamic viscosities up to 10 and 50 N s/m^2, respectively.

Figure (5.2–5) shows a turbine agitator of diameter D_A in a cylindrical tank of diameter D_T filled with liquid to a height H_L. The agitator is located at a height H_A from the bottom of the tank and the baffles which are located immediately adjacent to the wall have a width b. The agitator has a blade width a and blade length r and the blades are mounted on a central disc of diameter s. A typical turbine mixing system is the standard configuration defined by the following geometrical relationships:

(1) a 6-blade flat blade turbine agitator
(2) $D_A = D_T/3$
(3) $H_A = D_T/3$
(4) $a = D_T/5$
(5) $r = D_T/4$
(6) $H_L = D_T$
(7) 4 symmetrical baffles
(8) $b = D_T/10$

Processing considerations sometimes necessitate deviations from the standard configuration.

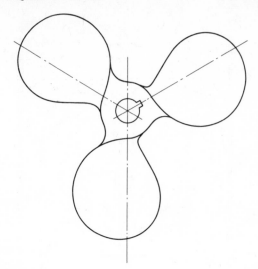

Figure (5.2–2)
Marine propeller.

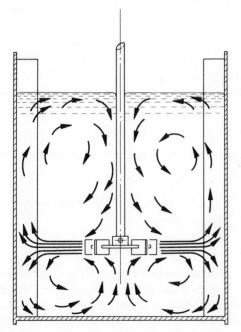

Figure (5.2–3)
Radial flow pattern produced by a flat blade turbine.

Figure (5.2–4)
Axial flow pattern produced by a marine propeller.

Agitator tip speeds TS given by equation (5.2–1) are commonly used as a measure of the degree of agitation in a liquid mixing system.

$$TS = \pi D_A N \qquad\qquad (5.2\text{–}1)$$

Tip speed ranges for turbine agitators are recommended as follows:

2.5 to 3.3 m/s for low agitation

3.3 to 4.1 m/s for medium agitation

and

4.1 to 5.6 m/s for high agitation

If turbine or marine propeller agitators are used to mix relatively low viscosity liquids in unbaffled tanks, vortexing develops. In this case the liquid level falls in the immediate vicinity of the agitator shaft. Vortexing increases with rotational speed N until eventually the vortex passes through the agitator. As the liquid viscosity increases, the need for baffles to reduce vortexing decreases.

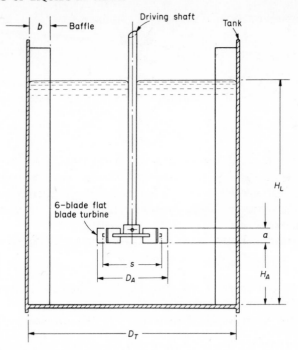

Figure (5.2–5)
Standard tank configuration.

A marine propeller can be considered as a caseless pump. In this case its volumetric circulating capacity Q_A is related to volumetric displacement per revolution V_D by the equation

$$Q_A = \eta V_D N \qquad (5.2–2)$$

where η is a dimensionless efficiency factor which is approximately 0.6.[22] V_D is related to the propeller pitch p and the propeller diameter D_A by the equation

$$V_D = \frac{\pi D_A^2 p}{4} \qquad (5.2–3)$$

Most propellers are square pitch propellers where $p = D_A$ so that equation (5.2–3) becomes

$$V_D = \frac{\pi D_A^3}{4} \qquad (5.2–4)$$

Combine equations (5.2–2) and (5.2–4) to give

$$Q_A = \frac{\eta \pi N D_A^3}{4} \qquad (5.2\text{–}5)$$

which is analogous to equation (4.4–4) for centrifugal pumps.

Weber[22] defined a tank turnover rate I_T by the equation

$$I_T = \frac{Q_A}{V} \qquad (5.2\text{–}6)$$

where V is the tank volume and I_T is the number of turnovers per unit time. To get the best mixing, I_T should be at a maximum. For a given tank volume V, this means that the circulating capacity Q_A should have the highest possible value for the minimum consumption of power.

The head developed by the rotating agitator h_A can be written as

$$h_A = C_1 N^2 D_A^2 \qquad (5.2\text{–}7)$$

where C_1 is a constant. Equation (5.2–7) is analogous to equation (4.4–5) for centrifugal pumps.

Combine equations (5.2–5) and (5.2–7) to give the ratio

$$\frac{Q_A}{h_A} = \frac{C D_A}{N} \qquad (5.2\text{–}8)$$

where C is a constant.

Since the mean shear rate in a mixing tank $\dot\gamma_m$ is given by equation (5.1–1)

$$\dot\gamma_m = kN \qquad (5.1\text{–}1)$$

equation (5.2–8) can also be written in the form

$$\frac{Q_A}{h_A} = \frac{C' D_A}{\dot\gamma_m} \qquad (5.2\text{–}9)$$

where C' is also a constant.

The ratio of circulating capacity to head Q_A/h_A is low for high shear agitators. For mixing pseudoplastic liquids a high circulating capacity Q_A and a high shear rate $\dot\gamma_m$ or head h_A are both desirable. In this case a compromise has to be made.

5.3   Large blade low speed agitators

Large blade low-speed agitators include anchors, gates, paddles, helical ribbons and helical screws. They are used to mix relatively high viscosity liquids and depend on a large blade area to produce liquid movement throughout a tank. Since they are low shear agitators they are useful for mixing dilatant liquids.

A gate type anchor agitator is shown in Figure (5.3–1). Anchor agitators operate within close proximity to the tank wall. The shearing action of the anchor blades past the tank wall produces a

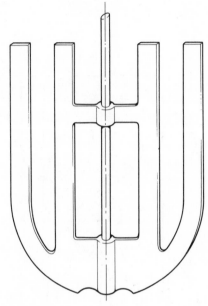

Figure (5.3–1)
Gate type anchor agitator.

continual interchange of liquid between the bulk liquid and the liquid film between the blades and the wall.[10] Anchors have successfully been used to mix liquids with dynamic viscosities up to 100 N s/m².[1][21] For heat transfer applications, anchors may be fitted with wall scrapers to prevent the build up of a stagnant film between the anchor and the tank wall.

Uhl and Voznick[21] showed that the mixing effectiveness of a particular anchor agitator in a Newtonian liquid of dynamic viscosity 40 N s/m² was the same as for a particular turbine agitator in a Newtonian liquid of dynamic viscosity 15 N s/m².

Helical screws normally function by pumping liquid from the bottom of a tank to the liquid surface. The liquid then returns to the bottom of the tank to fill the void created when fresh liquid is pumped to the surface. A rotating helical screw positioned vertically in the centre of an unbaffled cylindrical tank produces a mild swirling motion in the liquid. Since the liquid velocity decreases towards the tank wall, the liquid at the wall of an unbaffled tank is nearly motionless. Baffles set away from the tank wall create turbulence and facilitate the entrainment of liquid in contact with the tank wall.

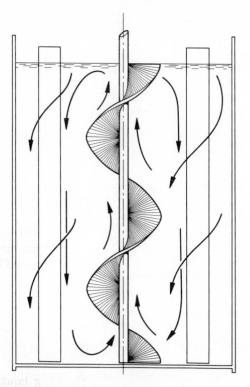

Figure (5.3–2)
Flow pattern in a baffled helical screw system.

The flow pattern in a baffled helical screw system is shown in Figure (5.3–2). Baffles are not required if the helical screw is placed in an off-centred position since in this case the system becomes self baffling. However, off-centred helical screws require more power to produce a comparable mixing result.

Gray[3] investigated the mixing times of helical ribbon agitators and found the following equation to hold:

$$Nt = 30 \qquad (5.3–1)$$

where N is the rotational speed of the helical ribbon agitator and t is the batch mixing time.

5.4 Dimensionless groups for mixing

In the design of liquid mixing systems the following dimensionless groups are of importance.

The power number

$$N_p = \frac{P_A}{\rho N^3 D_A^5} \qquad (5.4–1)$$

The Reynolds number for mixing $(N_{RE})_M$ represents the ratio of the applied to the opposing viscous drag forces.

$$(N_{RE})_M = \frac{\rho N D_A^2}{\mu} \qquad (5.4–2)$$

The Froude number for mixing $(N_{FR})_M$ represents the ratio of the applied to the opposing gravitational forces.

$$(N_{FR})_M = \frac{N^2 D_A}{g} \qquad (5.4–3)$$

The Weber number for mixing $(N_{WE})_M$ represents the ratio of the applied to the opposing surface tension forces.

$$(N_{WE})_M = \frac{\rho N^2 D_A^3}{\sigma} \qquad (5.4–4)$$

In the above equations, ρ, μ and σ are the density, dynamic viscosity and surface tension respectively of the liquid; P_A, N and D_A are the power consumption, rotational speed and diameter respectively of the agitator.

The terms in equations (5.4–1), (5.4–2), (5.4–3) and (5.4–4) must be in consistent units. In the SI system ρ is in kg/m³, μ in N s/m² and σ in N/m; P_A is in W, N in rev/s and D_A in m.

It can be shown by dimensional analysis[5] that the power number N_p can be related to the Reynolds number for mixing $(N_{RE})_M$ and the Froude number for mixing $(N_{FR})_M$ by the equation

$$N_p = C(N_{RE})_M^x (N_{FR})_M^y \qquad (5.4–5)$$

where C is an overall dimensionless shape factor which represents the geometry of the system.

Equation (5.4–5) can also be written in the form

$$\phi = \frac{N_p}{(N_{FR})_M^y} = C(N_{RE})_M^x \qquad (5.4\text{–}6)$$

where ϕ is defined as the dimensionless power function.

In liquid mixing systems, baffles are used to suppress vortexing. Since vortexing is a gravitational effect, the Froude number is not required to describe baffled liquid mixing systems. In this case the exponent y in equations (5.4–5) and (5.4–6) is zero and $(N_{FR})_M^y = 1$.

Thus for non-vortexing systems equation (5.4–6) can be written either as

$$\phi = N_p = C(N_{RE})_M^x \qquad (5.4\text{–}7)$$

or as

$$\log N_p = \log C + x \log (N_{RE})_M \qquad (5.4\text{–}8)$$

The Weber number for mixing $(N_{WE})_M$ is only of importance when separate physical phases are present in the liquid mixing system as in liquid–liquid extraction.

5.5   Power curves

A power curve is a plot of the power function ϕ or the power number N_p against the Reynolds number for mixing $(N_{RE})_M$ on log–log coordinates. Each geometrical configuration has its own power curve and since the plot involves dimensionless groups it is independent of tank size. Thus a power curve used to correlate power data in a $1\,m^3$ tank system is also valid for a $1000\,m^3$ tank system provided that both tank systems have the same geometrical configuration.

Figure (5.5–1) shows the power curve for the standard tank configuration geometrically illustrated in Figure (5.2–5). Since this is a baffled non-vortexing system equation (5.4–8) applies.

$$\log N_p = \log C + x \log (N_{RE})_M \qquad (5.4\text{–}8)$$

The power curve for the standard tank configuration is linear in the laminar flow region AB with a slope of -1.0. Thus in this region for $(N_{RE})_M < 10$, equation (5.4–8) can be written as

$$\log N_p = \log C - \log (N_{RE})_M \qquad (5.5\text{–}1)$$

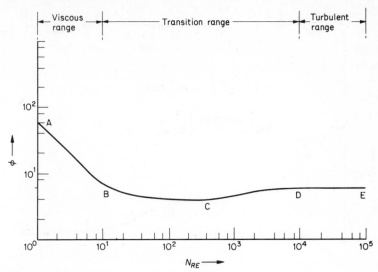

Figure (5.5–1)
Power curve for the standard tank configuration.

which can be rearranged to

$$P_A = \mu C N^2 D_A^3 \qquad (5.5\text{–}2)$$

where $C = 71.0$ for the standard tank configuration. Thus for laminar flow, power is directly proportional to dynamic viscosity for a fixed agitator speed.

For the transition flow region BCD which extends up to $(N_{RE})_M = 10\,000$, the parameters C and x in equation (5.4–8) vary continuously.

In the fully turbulent flow region DE, the curve becomes horizontal and the power function ϕ is independent of the Reynolds number for mixing $(N_{RE})_M$. For the region $(N_{RE})_M > 10\,000$

$$\phi = N_p = 6.3 \qquad (5.5\text{–}3)$$

At point C on the power curve for the standard tank configuration given in Figure (5.5–1) enough energy is being transferred to the liquid for vortexing to start. However the baffles in the tank prevent this. If the baffles were not present vortexing would develop and the power curve would be as shown in Figure (5.5–2).

The power curve in Figure (5.5–1) for the baffled system is identical with the power curve in Figure (5.5–2) for the unbaffled

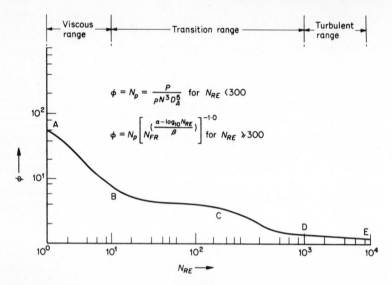

Figure (5.5–2)
Power curve for the standard tank configuration without baffles.

system up to point C where $(N_{RE})_M \cong 300$. As the Reynolds number for mixing $(N_{RE})_M$ increases beyond point C in the unbaffled system, vortexing increases and the power falls sharply.

Equation (5.4–5) can be written in the form

$$\log N_p = \log C + x \log (N_{RE})_M + y \log (N_{FR})_M \qquad (5.5\text{–}4)$$

For the unbaffled system, $\phi = N_p$ at $(N_{RE})_M < 300$ and $\phi = N_p/(N_{FR})_M^y$ at $(N_{RE})_M > 300$.

A plot of N_p against $(N_{RE})_M$ on log–log coordinates for the unbaffled system gives a family of curves at $(N_{RE})_M > 300$. Each curve has a constant Froude number for mixing $(N_{FR})_M$.

A plot of N_p against $(N_{FR})_M$ on log–log coordinates is a straight line of slope y at a constant Reynolds number for mixing $(N_{RE})_M$. A number of lines can be plotted for different values of $(N_{RE})_M$. A plot of y against $\log (N_{RE})_M$ is also a straight line. If the slope of the line is $-1/\beta$ and the intercept at $(N_{RE})_M = 1$ is α/β then

$$y = \frac{\alpha - \log (N_{RE})_M}{\beta} \qquad (5.5\text{–}5)$$

Substitute equation (5.5–5) into equation (5.4–6) to give

$$\phi = \frac{N_p}{(N_{FR})_M^{\{[\alpha - \log(N_{RE})_M]/\beta\}}}$$ (5.5–6)

Rushton, Costich, and Everett[20] have listed values of α and β for various vortexing systems. For a 6-blade flat blade turbine agitator 0.1 m in diameter $\alpha = 1.0$ and $\beta = 40.0$.

If a power curve is available for a particular system geometry, it can be used to calculate the power consumed by an agitator at various rotational speeds, liquid viscosities and densities. The procedure is as follows: calculate the Reynolds number for mixing $(N_{RE})_M$; read the power number N_p or the power function ϕ from the appropriate power curve and calculate the power P_A from either equation (5.4–1) rewritten in the form

$$P_A = N_p \rho N^3 D_A^5.$$ (5.5–7)

or equation (5.4–6) rewritten in the form

$$P_A = \phi \rho N^3 D_A^5 \left(\frac{N^2 D_A}{g}\right)^y$$ (5.5–8)

Equations (5.5–7) and (5.5–8) can be used to calculate only the power consumed by the agitator. Additional power is required to overcome electrical and mechanical losses which occur in all mixing systems.

The power curves given in Figures (5.5–1) and (5.5–2) were obtained from experiments using Newtonian liquids.

Magnusson[14] calculated the apparent viscosities of non-Newtonian liquids in agitated tanks from the appropriate power curves for Newtonian liquids. Metzner and Otto[15] used this procedure to obtain the dimensionless proportionality constant k in equation (5.1–1) and a non-Newtonian power curve for a particular system geometry.

$$\dot{\gamma}_m = kN$$ (5.1–1)

The procedure is as follows:

(1) Obtain power data using a non-Newtonian liquid and calculate the power number N_p from equation (5.4–1) for various agitator speeds N.

(2) Read the Reynolds number for mixing $(N_{RE})_M$ from the appropriate Newtonian power curve for each value of N_p and N.

(3) For each value of $(N_{RE})_M$ and N in the laminar flow region calculate the apparent viscosity μ_a from equation (5.4–2) rewritten in the form

$$\mu_a = \frac{\rho N D_A^2}{(N_{RE})_M} \qquad (5.5\text{–}9)$$

(4) Compare a log–log plot of μ_a against N with a log–log plot of μ_a against $\dot{\gamma}$ experimentally determined using a viscometer. Plot $\dot{\gamma}$ against N on ordinary Cartesian coordinates for corresponding values of μ_a. The plot is a straight line of slope k which is the dimensionless proportionality constant in equation (5.1–1).

(5) For various values of power number N_p and corresponding agitator speeds N beyond the laminar flow region calculate values of shear rate $\dot{\gamma}$ from equation (5.1–1). Read the corresponding values of apparent viscosity μ_a from the log–log plot of μ_a against $\dot{\gamma}$ and calculate the Reynolds number for mixing

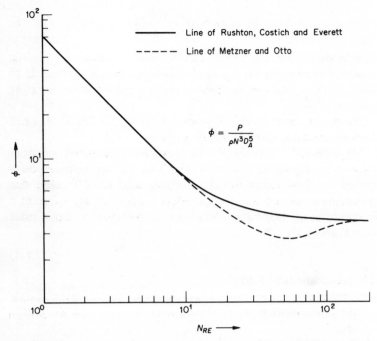

Figure (5.5–3)
Deviation from Newtonian power curve for pseudoplastic liquids.

$(N_{RE})_M$ for each value of N and N_p. Extend the power curve beyond the laminar flow region by plotting these values of N_p and $(N_{RE})_M$.

Figure (5.5–3) illustrates the use of this method to extend a Newtonian power curve in the laminar region into a non-Newtonian power curve. The full line is the Newtonian power curve obtained by Rushton, Costich and Everett[20] for a flat blade turbine system. The dashed line is a plot of the data of Metzner and Otto[15] for pseudoplastic liquids.

Figure (5.5–3) shows that at no point is the pseudoplastic power curve higher than the Newtonian power curve. Thus the use of the Newtonian power curve to calculate powers will give conservative values when used for pseudoplastic liquids. Figure (5.5–3) also shows that the laminar flow region for pseudoplastic liquids extends to higher Reynolds numbers than that for Newtonian liquids.

Example (5.5–1)

Calculate the theoretical power in W for a 3 m diameter, 6-blade flat blade turbine agitator running at 0.2 rev/s in a tank system conforming to the standard tank configuration illustrated in Figure (5.2–5). The liquid in the tank has a dynamic viscosity of 1.0 N s/m² and a liquid density of 1000 kg/m³.

Calculations:

Reynolds number for mixing

$$(N_{RE})_M = \frac{\rho N D_A^2}{\mu} \qquad (5.4\text{–}2)$$

$$\rho = 1000 \text{ kg/m}^3$$

$$N = 0.2 \text{ rev/s}$$

$$D_A = 3 \text{ m}$$

$$\mu = 1.0 \text{ N s/m}^2 = 1.0 \text{ kg/(s m)}$$

$$(N_{RE})_M = \frac{(1000 \text{ kg/m}^3)(0.2 \text{ rev/s})(9.0 \text{ m}^2)}{1.0 \text{ kg/(s m)}}$$

$$= 1800$$

from ϕ against $(N_{RE})_M$ graph in Figure (5.5–1)

$$\phi = N_p = 4.5$$

theoretical power for mixing

$$P_A = N_p \rho N^3 D_A^5 \tag{5.5-7}$$

$$= (4.5)(1000 \text{ kg/m}^3)(0.008 \text{ rev}^3/\text{s}^3)(243 \text{ m}^5)$$

$$= 8748 \text{ W}$$

Example (5.5–2)

Calculate the theoretical power in W for a 0.1 m diameter, 6-blade flat blade turbine agitator running at 16 rev/s in a tank system without baffles but otherwise conforming to the standard tank configuration illustrated in Figure (5.2–5). The liquid in the tank has a dynamic viscosity of 0.08 N s/m^2 and a liquid density of 900 kg/m^3.

Calculations:

Reynolds number for mixing

$$(N_{RE})_M = \frac{\rho N D_A^2}{\mu} \tag{5.4-2}$$

$$\rho = 900 \text{ kg/m}^3$$

$$N = 16 \text{ rev/s}$$

$$D_A = 0.1 \text{ m}$$

$$\mu = 0.08 \text{ N s/m}^2 = 0.08 \text{ kg/(s m)}$$

$$(N_{RE})_M = \frac{(900 \text{ kg/m}^3)(16 \text{ rev/s})(0.01 \text{ m}^2)}{0.08 \text{ kg/(s m)}}$$

$$= 1800$$

from ϕ against $(N_{RE})_M$ graph in Figure (5.5–2)

$$\phi = 2.2$$

theoretical power for mixing

$$P_A = \phi \rho N^3 D_A^5 \left(\frac{N^2 D_A}{g} \right)^y \tag{5.5-8}$$

$$y = \frac{\alpha - \log (N_{RE})_M}{\beta} \tag{5.5-5}$$

$$\alpha = 1.0, \qquad \beta = 40.0$$

$$\log 1800 = 3.2553$$

$$y = \frac{-2.2553}{40} = -0.05638$$

$$\frac{N^2 D_A}{g} = \frac{(256 \text{ rev}^2/\text{s}^2)(0.1 \text{ m})}{9.81 \text{ m/s}^2}$$

$$= 2.610$$

$$\left(\frac{N^2 D_A}{g}\right)^y = 2.610^{(-0.05638)} = 0.9479$$

$$P_A = (2.2)(900 \text{ kg/m}^3)(4096 \text{ rev}^3/\text{s}^3)(0.00001 \text{ m}^5)(0.9479)$$

$$= 76.88 \text{ W}$$

5.6   Scale-up of liquid mixing systems

The principle of similarity[4,12,16] together with the use of dimensionless groups is the essential basis of scale-up. The types of similarity relevant to liquid mixing systems together with their definitions are listed as follows:

Geometrical similarity exists between two systems of different sizes when the ratios of corresponding dimensions in one system are equal to those in the other. Hence, geometrical similarity exists between two pieces of equipment of different sizes when both have the same shape.

Kinematic similarity exists between two systems of different sizes when they are not only geometrically similar but when the ratios of velocities between corresponding points in one system are equal to those in the other.

Dynamic similarity exists between two systems when, in addition to being geometrically and kinematically similar, the ratios of forces between corresponding points in one system are equal to those in the other.

Dimensionless groups provide a convenient way of correlating scientific and engineering data.

The classical principle of similarity can be expressed by equations of the form

$$N_1 = f(N_2, N_3 \cdots) \qquad (5.6\text{--}1)$$

where a dimensionless group N_1 is a function of other dimensionless groups N_2, N_3, etc. Equation (5.6–1) is derived for a particular case by dimensional analysis, which is a technique for expressing the behaviour of a physical system in terms of the minimum number of independent variables.

Each dimensionless group represents a rule for scale-up. Frequently these individual scale-up rules conflict. For example, scale-up on dynamic similarity should depend chiefly upon a single dimensionless group that represents the ratio of the applied to the opposing forces. The Reynolds, Froude and Weber numbers are the ratios of the applied to the resisting viscous, gravitational and surfaces forces, respectively.

For scale-up from system 1 to system 2 for the same liquid properties and system geometry, equation (5.4–2) which defines the Reynolds number for mixing $(N_{RE})_M$, equation (5.4–3) which defines the Froude number for mixing $(N_{FR})_M$, and equation (5.4–4) which defines the Weber number for mixing $(N_{WE})_M$ can be written respectively in the following forms:

$$N_1 D_{A1}^2 = N_2 D_{A2}^2 \tag{5.6–2}$$

$$N_1^2 D_{A1} = N_2^2 D_{A2} \tag{5.6–3}$$

$$N_1^2 D_{A1}^3 = N_2^2 D_{A2}^3 \tag{5.6–4}$$

Clearly the scale-up rules represented by equations (5.6–2), (5.6–3) and (5.6–4) conflict. In order to scale up with accuracy, it is often necessary to design pilot equipment so that the effects of certain dimensionless groups are deliberately suppressed in favour of a particular dimensionless group.[7] For example, baffles can be used to eliminate vortexing so that the Froude number need not be considered.

Frequently it is not possible to achieve the desired similarity when scaling up from small to large scale units. In this case, results on the small scale unit must be extrapolated to dissimilar conditions on the large scale.

In order to extrapolate, use is made of what is known as the extended principle of similarity, where equations of the form

$$N_1 = C N_2^x N_3^y \ldots \tag{5.6–5}$$

are used. Here the dimensionless group N_1 is proportional to the dimensionless group N_2 to the xth power and the dimensionless group N_3 to the yth power etc. C is a constant that depends on the geometry of the system and is consequently a shape factor, which usually must be determined by experiment.

Equation (5.4–5) for liquid mixing systems is in the same form as equation (5.6–5)

$$N_p = C(N_{RE})_M^x (N_{FR})_M^y \qquad (5.4–5)$$

The scale-up of liquid mixing systems can be divided into two categories:[6] the scale-up of process result and the scale-up of power data.

The type of agitator and tank geometry required to achieve a particular process result, are determined from pilot plant experiments. The desired process result may be the dispersion or emulsification of immiscible liquids, the completion of a chemical reaction, the suspension of solids in a liquid or any one of a number of other processes.[8]

Once the process result has been satisfactorily obtained in the pilot size unit, it is necessary to predict the agitator speed in a geometrically similar production size unit using a suitable rule for scale-up.

Mutually conflicting scale-up rules are given by equations (5.6–2), (5.6–3) and (5.6–4). Other possible ways of scaling up are a constant tip speed TS, a constant ratio of circulating capacity to head Q_A/h_A, and a constant power per unit volume P_A/V. Since P_A is proportional to $N^3 D_A^5$ and V is proportional to D_A^3, the ratio P_A/V is proportional to $N^3 D_A^2$.

For scale-up from system 1 to system 2 for the same liquid properties and system geometry the following equations can be written:

$$N_1 D_{A1} = N_2 D_{A2} \qquad (5.6–6)$$

for a constant tip speed TS

$$\frac{D_{A1}}{N_1} = \frac{D_{A2}}{N_2} \qquad (5.6–7)$$

for a constant ratio of circulating capacity to head Q_A/h_A and

$$N_1^3 D_{A1}^2 = N_2^3 D_{A2}^2 \qquad (5.6–8)$$

for a constant power per unit volume P_A/V. The scale-up rules given by equations (5.6–2), (5.6–3), (5.6–4), (5.6–6), (5.6–7) and (5.6–8) are all mutually conflicting.

In practice, the process result and corresponding agitator speeds can be obtained in three small geometrically similar tank systems of different sizes. These data can then be extrapolated to give the agitator speed in a geometrically similar production size tank system which will give the desired process result.

The power curve obtained on a pilot size unit can be used directly to obtain the power requirements for a geometrically similar production size unit once the agitator speed is known.

Consider the scale up of the rotational speed of marine propellers for the same power consumption and Reynolds number for mixing.

The power consumption P_A is given by equation (5.5–7)

$$P_A = N_p \rho N^3 D_A^5 \qquad (5.5–7)$$

For scale-up from system 1 to system 2 for the same liquid properties and system geometry equation (5.5–7) can be written in the form

$$\frac{N_{p1} N_1^3 D_{A1}^5}{P_{A1}} = \frac{N_{p2} N_2^3 D_{A2}^5}{P_{A2}} \qquad (5.6–9)$$

which for the same power consumption and Reynolds number and hence power number becomes

$$N_1^3 D_{A1}^5 = N_2^3 D_{A2}^5 \qquad (5.6–10)$$

The corresponding equation for equality of Reynolds numbers for mixing has already been shown to be

$$N_1 D_{A1}^2 = N_2 D_{A2}^2 \qquad (5.6–2)$$

Divide equation (5.6–10) by equation (5.6–2) to give

$$N_1^2 D_{A1}^3 = N_2^2 D_{A2}^3 \qquad (5.6–11)$$

The circulating capacity Q_A of a square pitch propeller has already been shown to be

$$Q_A = \frac{\eta \pi N D_A^3}{4} \qquad (5.2–5)$$

For scale-up from system 1 to system 2, equation (5.2–5) can be written in the form

$$\frac{Q_{A1}}{\eta_1 N_1 D_{A1}^3} = \frac{Q_{A2}}{\eta_2 N_2 D_{A2}^3} \qquad (5.6–12)$$

Combine equations (5.6–11) and (5.6–12) to give

$$Q_{A2} = \left(\frac{\eta_2}{\eta_1}\right)\left(\frac{N_1}{N_2}\right) Q_{A1} \qquad (5.6–13)$$

Equation (5.6–13) shows that the circulation capacity of low-speed square pitch propellers greatly exceeds that of high speed propellers for the same power consumption and Reynolds number.[9]

5.7 The purging of stirred tank systems

In industry it is common practice to use a number of tanks equipped with agitators in series. Frequently it is necessary to know the time required to reduce the concentration of off-quality material in the system below a certain acceptable limit.

Let a mass m of solute be dissolved in a liquid volume V in a stirred tank. Let solute free liquid flow into the tank at a volumetric flow rate Q. Let liquid flow from the tank at a volumetric flow rate Q. If the liquid in the tank is uniformly mixed the discharge liquid contains a concentration of solute m/V which is the same as the solute concentration in the tank.

The rate of decrease of solute in the tank is given by the equation

$$\frac{\mathrm{d}m}{\mathrm{d}t} = -\frac{m}{V}Q \qquad (5.7–1)$$

which can be integrated to give

$$\frac{m}{m_0} = \frac{C_{1t}}{C_{10}} = \mathrm{e}^{-Qt/V} \qquad (5.7–2)$$

where C_{1t} is the solute concentration after time t and C_{10} is the initial solute concentration at time zero.

Equation (5.7–2) can also be written in the form

$$C_{1t} = C_{10}\,\mathrm{e}^{-\alpha t} \qquad (5.7–3)$$

where $\alpha = Q/V$ the reciprocal of the nominal holding time for the liquid in the tank.

The fraction x of the original solute which has been purged from the tank after a time t is given by the equation

$$x = \frac{m_0 - m}{m_0} = \frac{C_{10} - C_{1t}}{C_{10}} = 1 - \mathrm{e}^{-\alpha t} \qquad (5.7–4)$$

For a second tank of the same size in series, the rate of change of solute concentration is given by the equation

$$V\frac{\mathrm{d}C_{2t}}{\mathrm{d}t} = Q(C_{1t} - C_{2t}) \qquad (5.7–5)$$

which can be rewritten as

$$\frac{dC_{2t}}{dt} = \alpha(C_{1t} - C_{2t}) \qquad (5.7\text{--}6)$$

At time $t = 0$, $C_{2t} = 0$. For this case substitute equation (5.7–3) into equation (5.7–6) and integrate to give

$$C_{2t} = \alpha C_{10} t\, e^{-\alpha t} \qquad (5.7\text{--}7)$$

Similarly for a third tank of the same size in series, the rate of change of solute concentration is

$$\frac{dC_{3t}}{dt} = \alpha(C_{2t} - C_{3t}) \qquad (5.7\text{--}8)$$

At time $t = 0$, $C_{3t} = 0$. For this case substitute equation (5.7–7) into equation (5.7–8) and integrate to give

$$C_{3t} = \alpha^2 C_{10}\left(\frac{t^2}{2!}\right) e^{-\alpha t} \qquad (5.7\text{--}9)$$

Similarly, the concentration of solute in an nth tank of the same size in series can be written as

$$C_{nt} = \alpha^{n-1} C_{10}\left[\frac{t^{n-1}}{(n-1)!}\right] e^{-\alpha t} \qquad (5.7\text{--}10)$$

where at time $t = 0$, $C_{nt} = 0$. Add equations (5.7–3), (5.7–7), (5.7–9) and (5.7–10) to give the equation

$$V(C_{1t} + C_{2t} + C_{3t} + \ldots + C_{nt})$$
$$\qquad (5.7\text{--}11)$$
$$= VC_{10}\, e^{-\alpha t}\left[1 + \alpha t + \frac{\alpha^2 t^2}{2!} + \ldots + \frac{\alpha^{n-1} t^{n-1}}{(n-1)!}\right]$$

Equation (5.7–11) gives the total amount of solute remaining in a system of n equal size tanks after a time t where at time $t = 0$ the concentration in the first tank was C_{10} and the concentration in all other tanks was zero.

The amount of solute purged from the system after time t is given by the equation

$$m_t = V\left\{C_{10} - C_{10}\, e^{-\alpha t}\left[1 + \alpha t + \frac{\alpha^2 t^2}{2!} + \ldots + \frac{\alpha^{n-1} t^{n-1}}{(n-1)!}\right]\right\}$$
$$\qquad (5.7\text{--}12)$$

The fraction x of the original solute which has been purged from the system after time t is

$$x = 1 - e^{-\alpha t}\left[1 + \alpha t + \frac{\alpha^2 t^2}{2!} + \ldots + \frac{\alpha^{n-1} t^{n-1}}{(n-1)!}\right] \qquad (5.7\text{--}13)$$

Equation (5.7–13) is known as the purging time equation for a system of continuous mixing vessels of equal size in series. It has been derived by a number of investigators.[2,11,13]

REFERENCES

(1) Brown, R. W., Scott, R., and Toyne, C., *Transactions of the Institute of Chemical Engineers*, **25**, 181 (1947).
(2) Denbigh, K. G., Trans. Faraday Society, **40**, 352 (1944).
(3) Gray, J. B., A.I.Ch.E. Symposium Series Reprint No. 18 (1962).
(4) Holland, F. A., Chem. and Proc. Eng., **45**, 121 (1964).
(5) Holland, F. A., and Chapman, F. S., *Liquid Mixing and Processing in Stirred Tanks*, p. 4, New York, Reinhold Publishing Corporation, 1966.
(6) Ibid., p. 50.
(7) Ibid., p. 51.
(8) Ibid., p. 55.
(9) Ibid., p. 86.
(10) Ibid., p. 87.
(11) Ibid., p. 112.
(12) Johnstone, R. E., and Thring, M. W., *Pilot Plants, Models and Scale-up Methods in Chemical Engineering*, p. 63, New York, McGraw-Hill Book Co. Inc., 1957.
(13) McMullin, R. B., and Weber, M., Jr., *Transactions of the American Institute of Chemical Engineers*, **31**, 409 (1935).
(14) Magnusson, K., Iva (Sweden), **23**, 86 (1952).
(15) Metzner, A. B., and Otto, R. E., A.I.Ch.E. Journal, **3**, 3 (1957).
(16) Newton, I., Phil. Trans. Royal Soc. (London), **22**, 824 (1701).
(17) Norwood, K. W., and Metzner, A. B., A.I.Ch.E. Journal, **6**, 432 (1960).
(18) Parker, N. H., Chem. Eng., **71**, No. 12 (1964).
(19) Quillen, C. S., Chem. Eng., **61**, No. 6 (1954).
(20) Rushton, J. H., Costich, E. W., and Everett, H. J., Chem. Eng. Progr., **46**, 395 (1950).
(21) Uhl, V. W., and Voznick, H. P., Chem. Eng. Progr., **56**, 72 (1960).
(22) Weber, A. P., Chem. Eng., **70**, No. 18 (1963).

6
Flow of compressible fluids in conduits

6.1 Energy relationships

The total energy E per unit mass of fluid is given by either of the following equations:

$$E = U + zg + \frac{P}{\rho} + \frac{v^2}{2} \qquad (1.6\text{--}1)$$

or

$$E = U + zg + PV + \frac{v^2}{2} \qquad (6.1\text{--}1)$$

where U, zg, P/ρ and $v^2/2$ are the internal, potential, pressure and kinetic energies per unit mass respectively and V is the volume of unit mass of fluid.

Consider unit mass of fluid flowing in steady state from a point 1 to a point 2. Between these two points, let a net amount of heat energy Δq be added to the fluid and let a net amount of work ΔW be done on the fluid.

An energy balance for unit mass of fluid can be written either as

$$E_1 + \Delta q + \Delta W = E_2 \qquad (6.1\text{--}2)$$

or as

$$(U_2 - U_1) + (z_2 - z_1)g + (P_2V_2 - P_1V_1) + \frac{(v_2^2 - v_1^2)}{2} = \Delta q + \Delta W$$

$$(6.1\text{--}3)$$

For steady flow in a pipe or tube the kinetic energy term can be written as $u^2/(2\alpha)$ where u is the mean linear velocity in the pipe or tube and α is a dimensionless correction factor which accounts for the velocity distribution across the pipe or tube. Fluids which are treated as compressible are almost always in turbulent flow and for a pipe of circular cross-section α is approximately 1 for turbulent flow.

Thus for a compressible fluid flowing in a pipe or tube, equation (6.1–3) can be written as

$$(U_2 - U_1) + (z_2 - z_1)g + (P_2V_2 - P_1V_1) + \frac{(u_2^2 - u_1^2)}{2} = \Delta q + \Delta W$$

(6.1–4)

where in SI units each term is in J/kg.

Since the enthalpy per unit mass of a fluid H is defined by the equation

$$H = U + PV \tag{6.1–5}$$

equation (6.1–4) can be written in the alternative form

$$(H_2 - H_1) + (z_2 - z_1)g + \frac{(u_2^2 - u_1^2)}{2} = \Delta q + \Delta W \tag{6.1–6}$$

where in SI units each term is also in J/kg.

The work term ΔW in equations (6.1–5) and (6.1–6) is positive if work is done on the fluid by a pump or compressor. ΔW is negative if the fluid does work in a turbine. ΔW is often referred to as shaft work since it is transmitted into or out of a system by means of a shaft.

Differentiate equation (6.1–4) to give

$$dU + g\,dz + P\,dV + V\,dP + d\left(\frac{u^2}{2}\right) = dq + dW \tag{6.1–7}$$

The first law of thermodynamics can be expressed by the equation

$$dq = dU + P\,dV \tag{6.1–8}$$

where dU is the increase in internal energy per unit mass of fluid and $P\,dV$ is the work of expansion on the fluid layers ahead for a net addition of heat per unit mass dq to the system.

Energy is required to overcome friction in a system. The effect of friction is to generate heat in a system by converting mechanical to thermal energy. Thus where friction is involved equation (6.1–8) can be written either as

$$dq + dF = dU + P\,dV$$

or as

$$dq = dU + P\,dV - dF \qquad (6.1–9)$$

where dF is the energy per unit mass required to overcome friction in the system.

Substitute equation (6.1–9) into equation (6.1–7) and write

$$g\,dz + V\,dP + d\left(\frac{u^2}{2}\right) + dF = dW \qquad (6.1–10)$$

Integrate equation (6.1–10) to give

$$(z_2 - z_1)g + \int_1^2 V\,dP + \frac{(u_2^2 - u_1^2)}{2} + \Delta F = \Delta W \quad (6.1–11)$$

where in SI units each term is in J/kg.

Equations (6.1–4), (6.1–6) and (6.1–11) all relate the energies involved for a fluid flowing in steady turbulent flow through a pipe of circular cross-section. The most appropriate equation is used in any particular application.

For incompressible fluids of density ρ, equation (6.1–11) can be written in the form

$$\left(z_2 + \frac{P_2}{\rho_2 g} + \frac{u_2^2}{2g}\right) - \left(z_1 + \frac{P_1}{\rho_1 g} + \frac{u_1^2}{2g}\right) = \frac{\Delta W}{g} - \frac{\Delta F}{g} \qquad (6.1–12)$$

where each term has the dimensions of length. Equation (6.1–12) is equivalent to equation (1.6–7) for the steady turbulent flow of an incompressible fluid through a pipe of circular cross-section.

It is sometimes more convenient to write equation (6.1–10) in terms of the mass flow rate of a fluid in a pipe G where $G = \rho u = u/V$. In SI units G is in kg/(s m²), ρ is in kg/m³, u is in m/s and V is in m³/kg.

In terms of G write equation (6.1–10) as

$$\rho^2 g\,dz + \frac{dP}{V} + \frac{G^2\,d}{V^2}\left(\frac{V^2}{2}\right) + \rho^2\,dF = \rho^2\,dW \qquad (6.1–13)$$

Integrate equation (6.1–13) to give either

$$\int_1^2 \rho^2 g \, dz + \int_1^2 \frac{dP}{V} + G^2 \ln\left(\frac{V_2}{V_1}\right) + \int_1^2 \rho^2 \, dF = \int_1^2 \rho^2 \, dW$$

$$(6.1–14)$$

or

$$\int_1^2 \rho^2 g \, dz + \int_1^2 \rho \, dP + G^2 \ln\left(\frac{\rho_1}{\rho_2}\right) + \int_1^2 \rho^2 \, dF = \int_1^2 \rho^2 \, dW$$

$$(6.1–15)$$

The pressure drop in a pipe of length L and inside diameter d_i is given by equation (2.4–3)

$$\Delta P = 8 j_f \left(\frac{L}{d_i}\right) \frac{\rho u^2}{2} \qquad (2.4–3)$$

where j_f is the dimensionless basic friction factor.

For a small element of pipe dL equation (2.4–3) can be written as

$$dP = 8 j_f \left(\frac{dL}{d_i}\right) \frac{\rho u^2}{2} \qquad (6.1–16)$$

The corresponding energy required to overcome friction $dF = dP/\rho$. Thus equation (6.1–16) can be written either as

$$dF = 8 j_f \left(\frac{dL}{d_i}\right) \frac{u^2}{2} \qquad (6.1–17)$$

or in terms of the mass flow rate G as

$$dF = 4 j_f \left(\frac{dL}{d_i}\right) \frac{G^2}{\rho^2} \qquad (6.1–18)$$

Substitute equation (6.1–18) into equations (6.1–14) and (6.1–15) respectively and write

$$\int_1^2 \rho^2 g \, dz + \int_1^2 \frac{dP}{V} + G^2 \ln\left(\frac{V_2}{V_1}\right) + \int_1^2 4 j_f \left(\frac{dL}{d_i}\right) G^2 = \int_1^2 \rho^2 \, dW$$

$$(6.1–19)$$

and

$$\int_1^2 \rho^2 g \, dz + \int_1^2 \rho \, dP + G^2 \ln\left(\frac{\rho_1}{\rho_2}\right) + \int_1^2 4 j_f \left(\frac{dL}{d_i}\right) G^2 = \int_1^2 \rho^2 \, dW$$

$$(6.1–20)$$

In order to make use of equations (6.1–19) and (6.1–20) it is necessary to know the relationship between pressure P and volume per unit mass V for a particular fluid before the term $\int_1^2 dP/V$ or $\int_1^2 \rho\, dP$ can be integrated. The relationship between P and V is given in an equation of state.

6.2   Equations of state

For ideal or perfect gases at a constant temperature the equation of state is

$$PV = \text{constant} \qquad (6.2\text{--}1)$$

Equation (6.2–1) is known as Boyle's law and can either be written as

$$P_1 V_1 = P_2 V_2 \qquad (6.2\text{--}2)$$

or in the differential form as

$$P\, dV + V\, dP = 0 \qquad (6.2\text{--}3)$$

The equation

$$PV = \frac{R_G T}{(MW)} \qquad (6.2\text{--}4)$$

also holds for ideal gases where R_G is the gas constant, T is the absolute temperature and (MW) is the molecular weight. Equation (6.2–4) is a combination of Boyle's and Charles' laws. In SI units $R_G = 8.3143$ kJ/(kmol K) and T is in K.

Many gases obey equation (6.2–4) reasonably well at ordinary temperatures and at low pressures up to and even beyond atmospheric pressure. At high pressures equation (6.2–4) is written in the modified form

$$PV = \frac{N_C R_G T}{(MW)} \qquad (6.2\text{--}5)$$

where N_C is the dimensionless compressibility factor. N_C is a function of the reduced pressure P_r and the reduced temperature T_r. P_r is the ratio of the actual pressure P to the critical pressure P_c.

$$P_r = \frac{P}{P_c} \qquad (6.2\text{--}6)$$

T_r is the ratio of the actual temperature T to the critical temperature T_c.

$$T_r = \frac{T}{T_c} \tag{6.2--7}$$

Plots of N_C against P_r at constant T_r are available.[4]

When ideal gases are compressed or expanded they obey the following equation:

$$PV^k = \text{constant} \tag{6.2--8}$$

Equation (6.2–8) can also be written either as

$$P_1 V_1^k = P_2 V_2^k \tag{6.2--9}$$

or as

$$\frac{P_2}{P_1} = \left(\frac{V_2}{V_1}\right)^k \tag{6.2--10}$$

Equation (6.2–10) can also be combined with equation (6.2–4) and written in the form

$$\frac{P_2}{P_1} = \left(\frac{T_2}{T_1}\right)^{k/(k-1)} \tag{6.2--11}$$

A change of state according to equation (6.2–8) is called a polytropic change.

For isothermal compression or expansion $k = 1$ and equation (6.2–9) becomes identical with equation (6.2–2).

For reversible adiabatic compression or expansion, i.e. when no heat transfer is allowed into or out of the fluid, $k = \gamma$ where $\gamma = C_p/C_v$ the ratio of the heat capacity per unit mass at constant pressure C_p to the heat capacity per unit mass at constant volume C_v. For the reversible adiabatic change of state of an ideal gas, equation (6.2–8) becomes

$$PV^\gamma = \text{constant} \tag{6.2--12}$$

and equation (6.2–11) becomes

$$\frac{P_2}{P_1} = \left(\frac{T_2}{T_1}\right)^{\gamma/(\gamma-1)} \tag{6.2--13}$$

Typical values of γ for ordinary temperatures and pressure are 1.67 for monatomic gases such as helium and argon, 1.40 for diatomic gases such as hydrogen, carbon monoxide and nitrogen, and 1.30 for triatomic gases such as carbon dioxide.

6.3   Sonic velocity in fluids

The sonic velocity u_s is the velocity of propagation of a pressure wave in a fluid. It can be shown[5] that u_s is related to the compressibility of the fluid $d\rho/dP$ by equation (6.3–1).

$$u_s = \sqrt{\frac{dP}{d\rho}} \tag{6.3–1}$$

The density of the fluid

$$\rho = \frac{1}{V} \tag{6.3–2}$$

Differentiate equation (6.3–2) to give

$$\frac{d\rho}{dV} = -\frac{1}{V^2} \tag{6.3–3}$$

and rewrite in the form

$$\frac{dP}{d\rho} = -V^2 \frac{dP}{dV} \tag{6.3–4}$$

When ideal gases are compressed or expanded they obey equation (6.2–8).

$$PV^k = \text{constant} \tag{6.2–8}$$

In equation (6.2–8), $k = 1$ for an isothermal change of state and $k = \gamma$ for a reversible adiabatic change of state where $\gamma = C_p/C_v$.

Differentiate equation (6.2–8) to give

$$\frac{dP}{dV} = -k\frac{P}{V} \tag{6.3–5}$$

Combine equations (6.3–4) and (6.3–5) to give

$$\frac{dP}{d\rho} = kPV \tag{6.3–6}$$

Substitute equation (6.3–6) into equation (6.3–1) to give

$$u_s = \sqrt{kPV} \tag{6.3–7}$$

where u_s is the sonic velocity or velocity of sound in the fluid.

For isothermal conditions equation (6.3–7) becomes

$$u_s = \sqrt{PV} \qquad (6.3\text{–}8)$$

For adiabatic conditions equation (6.3–7) becomes

$$u_s = \sqrt{\gamma PV} \qquad (6.3\text{–}9)$$

It is common practice to state gas velocities relative to the sonic velocity. The equation

$$N_M = \frac{u}{u_s} \qquad (6.3\text{–}10)$$

defines the dimensionless Mach number N_M. When $N_M > 1$ the velocity is supersonic.

6.4 Isothermal flow of an ideal gas in a horizontal pipe

For steady flow of an ideal gas between points 1 and 2 distance L apart in a horizontal pipe of constant circular cross-section in which no shaft work is done, equation (6.1–19) can be written as

$$\int_1^2 \frac{\mathrm{d}P}{V} + G^2 \ln\left(\frac{V_2}{V_1}\right) + 4j_f\left(\frac{L}{d_i}\right)G^2 = 0 \qquad (6.4\text{–}1)$$

where the dimensionless friction factor j_f is assumed to be substantially constant.

For an isothermal change of state of an ideal gas

$$PV = P_1 V_1 \qquad (6.4\text{–}2)$$

and

$$\int_1^2 \frac{\mathrm{d}P}{V} = \frac{1}{P_1 V_1} \int_1^2 P \, \mathrm{d}P \qquad (6.4\text{–}3)$$

which becomes

$$\int_1^2 \frac{\mathrm{d}P}{V} = \frac{(P_2^2 - P_1^2)}{2P_1 V_1} \qquad (6.4\text{–}4)$$

Combine equation (6.2–2) with the second term in equation (6.4–1) and write

$$G^2 \ln\left(\frac{V_2}{V_1}\right) = G^2 \ln\left(\frac{P_1}{P_2}\right) \qquad (6.4\text{–}5)$$

Substitute equations (6.4–4) and (6.4–5) into equation (6.4–1) and write

$$\frac{(P_2^2 - P_1^2)}{2P_1V_1} + G^2 \ln\left(\frac{P_1}{P_2}\right) + 4j_f\left(\frac{L}{d_i}\right)G^2 = 0 \qquad (6.4\text{–}6)$$

$$\frac{(P_2 + P_1)}{2}V_m = P_1V_1 \qquad (6.4\text{–}7)$$

where V_m is the mean volume per unit mass of gas in the pipe. Substitute equation (6.4–7) into equation (6.4–6) and write

$$\frac{(P_2 - P_1)}{V_m} + G^2 \ln\left(\frac{P_1}{P_2}\right) + 4j_f\left(\frac{L}{d_i}\right)G^2 = 0 \qquad (6.4\text{–}8)$$

If $(P_1 - P_2)/P_1 < 0.1$, the term $G^2 \ln(P_1/P_2)$ is negligible and equation (6.4–8) can be written as

$$P_1 - P_2 = 4j_f\left(\frac{L}{d_i}\right)G^2 V_m \qquad (6.4\text{–}9)$$

Since $G = \rho u$ and $V = 1/\rho$, equation (6.4–9) can be written as

$$P_1 - P_2 = 8j_f\left(\frac{L}{d_i}\right)\frac{\rho_m u_m^2}{2} \qquad (6.4\text{–}10)$$

Equation (6.4–10) is in the same form as equation (2.4–3) for the pressure drop of an incompressible fluid flowing in steady state through a pipe of circular cross-section.

$$\Delta P = 8j_f\left(\frac{L}{d_i}\right)\frac{\rho u^2}{2} \qquad (2.4\text{–}3)$$

Thus for small pressure drops, gases flowing through pipes can be treated as incompressible fluids. For large pressure drops, it is necessary to use equation (6.4–6).

The mass flow rate $G = 0$ when $P_1 = P_2$ and when $P_2 = 0$. At some intermediate pressure P_2, the mass flow rate G reaches a maximum.

Rewrite equation (6.4–6) in the form

$$\frac{(P_2^2 - P_1^2)}{2P_1V_1G^2} + \ln\left(\frac{P_1}{P_2}\right) + 4j_f\left(\frac{L}{d_i}\right) = 0 \qquad (6.4\text{–}11)$$

Differentiate equation (6.4–11) with respect to P_2 for a constant pressure P_1 to give

$$\frac{P_2}{P_1 V_1 G^2} - \frac{2}{G^3}\left(\frac{dG}{dP_2}\right)\left(\frac{P_2^2 - P_1^2}{2P_1 V_1}\right) - \frac{1}{P_2} = 0 \qquad (6.4\text{–}12)$$

The mass flow rate G is a maximum when $dG/dP_2 = 0$. At this maximum mass flow rate G_s, let the pressure $P_2 = P_s$, the mean linear velocity $u = u_s$, the density $\rho = \rho_s$, and the volume per unit mass $V = V_s$. For this case equation (6.4–12) can be written as

$$G_s^2 = \frac{P_s^2}{P_1 V_1} \qquad (6.4\text{–}13)$$

Since the maximum mass flow rate $G_s = \rho_s u_s = u_s/V_s$ and for isothermal conditions

$$P_1 V_1 = P_s V_s \qquad (6.4\text{–}14)$$

equation (6.4–13) can be rewritten in the form

$$u_s = \sqrt{P_s V_s} \qquad (6.4\text{–}15)$$

Since for isothermal conditions

$$PV = P_s V_s \qquad (6.4\text{–}16)$$

combine equations (6.4–15) and (6.4–16) to give

$$u_s = \sqrt{PV} \qquad (6.3\text{–}8)$$

It has already been shown that equation (6.3–8) gives the sonic velocity for isothermal conditions. Thus the maximum possible mean linear velocity of a gas in steady isothermal flow through a horizontal pipe of constant circular cross-section is the sonic velocity.

Supersonic velocities cannot be attained in a pipe of constant cross-sectional flow area. If the pressure P_2 is gradually reduced from an initial value $P_2 = P_1$, the mass flow rate will gradually increase until it reaches a maximum value G_s at a pressure $P_2 = P_s$. As the pressure P_2 is further reduced the mass flow rate remains constant at the maximum value G_s.

6.5 Non-isothermal flow of an ideal gas in a horizontal pipe

For steady flow of an ideal gas between points 1 and 2 distance L apart in a horizontal pipe of constant circular cross-section in which no shaft work is done, equation (6.1–19) can be written as

$$\int_1^2 \frac{dP}{V} + G^2 \ln\left(\frac{V_2}{V_1}\right) + 4j_f\left(\frac{L}{d_i}\right)G^2 = 0 \qquad (6.4\text{–}1)$$

where the dimensionless friction factor j_f is assumed to be substantially constant.

For a non-isothermal change of state of an ideal gas, equation (6.2–8) holds.

$$PV^k = \text{constant} \qquad (6.2\text{–}8)$$

Rewrite equation (6.2–8) in the form

$$PV^k = P_1 V_1^k \qquad (6.5\text{–}1)$$

and then in the form

$$\frac{1}{V} = \frac{P^{1/k}}{P_1^{1/k}V_1} \qquad (6.5\text{–}2)$$

Substitute equation (6.5–2) into the first term of equation (6.4–1) to give

$$\int_1^2 \frac{dP}{V} = \frac{1}{P_1^{1/k}V_1}\int_1^2 P^{1/k}\,dP \qquad (6.5\text{–}3)$$

Carry out the integration and write

$$\int_1^2 \frac{dP}{V} = \left(\frac{k}{k+1}\right)\left(\frac{P_1}{V_1}\right)\left[\left(\frac{P_2}{P_1}\right)^{(k+1)/k} - 1\right] \qquad (6.5\text{–}4)$$

Rewrite equation (6.2–8) in the form

$$\ln\left(\frac{V_2}{V_1}\right) = \frac{1}{k}\ln\left(\frac{P_1}{P_2}\right) \qquad (6.5\text{–}5)$$

Substitute equations (6.5–4) and (6.5–5) into equation (6.4–1) to give

$$\left(\frac{k}{k+1}\right)\left(\frac{P_1}{V_1}\right)\left[\left(\frac{P_2}{P_1}\right)^{(k+1)/k} - 1\right] + \frac{G^2}{k}\ln\left(\frac{P_1}{P_2}\right) + 4j_f\left(\frac{L}{d_i}\right)G^2 = 0$$
$$(6.5\text{–}6)$$

Equation (6.5–6) reduces to equation (6.4–6) for the isothermal case when $k = 1$.

If the pressure P_2 is gradually reduced from an initial value $P_2 = P_1$, the mass flow rate will gradually increase until it reaches a maximum value G_s at a pressure $P_2 = P_s$. At this point it can be shown[3] that the mean linear velocity u_s is given by the equation

$$u_s = \sqrt{kP_sV_s} \qquad (6.5\text{-}7)$$

Equation (6.5–7) is equivalent to equation (6.3–7)

$$u_s = \sqrt{kPV} \qquad (6.3\text{-}7)$$

It has already been shown that equation (6.3–7) gives the sonic velocity for non-isothermal conditions. Thus the maximum possible mean linear velocity of a gas in steady non-isothermal flow through a horizontal pipe of constant circular cross-section is the sonic velocity.

Equation (6.5–6) is only approximately true since in practice, k is not constant over the whole length of the pipe.

6.6 Adiabatic flow of an ideal gas in a horizontal pipe

Equation (6.1–6) relates the energies in a compressible fluid flowing in steady turbulent flow through a pipe of constant circular cross-section

$$(H_2 - H_1) + (z_2 - z_1)g + \frac{(u_2^2 - u_1^2)}{2} = \Delta q + \Delta W \quad (6.1\text{-}6)$$

For adiabatic flow between points 1 and 2 in a horizontal pipe in which no shaft work is done, equation (6.1–6) can be written as

$$(H_2 - H_1) + \frac{(u_2^2 - u_1^2)}{2} = 0 \qquad (6.6\text{-}1)$$

The enthalpy per unit mass of a fluid H is defined by equation (6.1–5)

$$H = U + PV \qquad (6.1\text{-}5)$$

For an ideal gas, the following thermodynamic relationships hold[2]:

$$U = C_v T \qquad (6.6\text{-}2)$$

$$C_p - C_v = \frac{R_G}{(MW)} \qquad (6.6\text{-}3)$$

$$PV = \frac{R_G T}{(MW)} \qquad (6.2\text{-}4)$$

Combine equations (6.1–5), (6.2–4), (6.6–2) and (6.6–3) to give

$$H = C_p T \qquad\qquad (6.6\text{–}4)$$

Substitute equation (6.6–4) into equation (6.6–1) and write

$$C_p(T_2 - T_1) + \frac{(u_2^2 - u_1^2)}{2} = 0 \qquad\qquad (6.6\text{–}5)$$

Equation (6.6–5) shows that an increase in gas velocity is accompanied by a decrease in temperature.

If the temperature at zero gas velocity is T_o, equation (6.6–5) may be written either as

$$C_p(T - T_o) + \frac{u^2}{2} = 0 \qquad\qquad (6.6\text{–}6)$$

or as

$$T = T_o - \frac{u^2}{2C_p} \qquad\qquad (6.6\text{–}7)$$

where T_o is known as the stagnation temperature. A thermometer placed in a flowing gas indicates a temperature T which is lower than the static or stagnation temperature T_o which can only be directly measured by a thermometer moving with the same velocity as the gas.

In practice equation (6.6–7) is used in the modified form

$$T = T_o - \frac{r_f u^2}{2C_p} \qquad\qquad (6.6\text{–}8)$$

where r_f is the recovery factor. For thermometers of conventional design $r_f = 0.88$[1].

Combine equations (6.2–4) and (6.6–3) to give

$$(C_p - C_v)T = PV \qquad\qquad (6.6\text{–}9)$$

Divide equation (6.6–9) by C_v to give

$$(\gamma - 1)T = \frac{PV}{C_v} \qquad\qquad (6.6\text{–}10)$$

in terms of the ratio $\gamma = C_p/C_v$. Multiply equation (6.6–10) by C_p and rewrite in the form

$$C_p T = \left(\frac{\gamma}{\gamma - 1}\right) PV \qquad\qquad (6.6\text{–}11)$$

Substitute equation (6.6–11) into equation (6.6–5) to give

$$\left(\frac{\gamma}{\gamma - 1}\right)(P_2 V_2 - P_1 V_1) + \frac{(u_2^2 - u_1^2)}{2} = 0 \qquad (6.6\text{–}12)$$

which can be rewritten in the form

$$P_2 V_2 \left[1 + \frac{(\gamma - 1)u_2^2}{2\gamma P_2 V_2}\right] - P_1 V_1 \left[1 + \frac{(\gamma - 1)u_1^2}{2\gamma P_1 V_1}\right] = 0 \quad (6.6\text{–}13)$$

Equation (6.3–9) gives the sonic velocity u_s of an ideal gas for adiabatic conditions.

$$u_s = \sqrt{\gamma P V} \qquad (6.3\text{–}9)$$

Equation (6.3–10) defines the dimensionless Mach number N_M.

$$N_M = \frac{u}{u_s} \qquad (6.3\text{–}10)$$

Combine equations (6.3–9), (6.3–10) and (6.6–13) to give

$$P_2 V_2 \left[1 + \frac{(\gamma - 1)}{2} N_{M2}^2\right] - P_1 V_1 \left[1 + \frac{(\gamma - 1)}{2} N_{M1}^2\right] = 0 \qquad (6.6\text{–}14)$$

where N_{M1} and N_{M2} are the dimensionless Mach numbers at points 1 and 2 respectively in a horizontal pipe of constant circular cross-section.

The mass flow rate $G = \rho u = u/V$. Therefore

$$u = GV \qquad (6.6\text{–}15)$$

Substitute equation (6.6–15) into equation (6.6–13) and write in the more general form

$$PV \left[1 + \frac{(\gamma - 1)G^2 V^2}{2\gamma PV}\right] = P_1 V_1 \left[1 + \frac{(\gamma - 1)G^2 V_1^2}{2\gamma P_1 V_1}\right] \qquad (6.6\text{–}16)$$

which can be rewritten as

$$PV = \frac{2\gamma P_1 V_1 + (\gamma - 1)G^2 V_1^2}{2\gamma} - \frac{(\gamma - 1)G^2 V^2}{2\gamma} \qquad (6.6\text{–}17)$$

Differentiate the product PV to give

$$d(PV) = V\,dP + P\,dV$$

and rewrite in the form

$$V\,dP = d(PV) - PV\frac{dV}{V} \qquad (6.6\text{--}18)$$

Combine equations (6.6–5), (6.6–11) and (6.6–15) and rewrite in the differential form as

$$d(PV) = -\left(\frac{\gamma - 1}{\gamma}\right)G^2 V\,dV \qquad (6.6\text{--}19)$$

It has already been shown that the energies in a compressible fluid flowing in a pipe are related by equation (6.1–10).

$$g\,dz + V\,dP + d\left(\frac{u^2}{2}\right) + dF = dW \qquad (6.1\text{--}10)$$

For flow in a horizontal pipe with no shaft work, equation (6.1–10) can be written as

$$V\,dP + u\,du + dF = 0 \qquad (6.6\text{--}20)$$

where the energy required to overcome friction dF is given by equation (6.1–17).

$$dF = 8j_f\left(\frac{dL}{d_i}\right)\frac{u^2}{2} \qquad (6.1\text{--}17)$$

Combine equations (6.1–17), (6.6–15) and (6.6–20) to give

$$V\,dP + G^2 V\,dV + 8j_f\left(\frac{dL}{d_i}\right)\frac{G^2 V^2}{2} = 0 \qquad (6.6\text{--}21)$$

Combine equations (6.6–17), (6.6–18) and (6.6–19), substitute into equation (6.6–21) and rearrange to give

$$\left(\frac{\gamma + 1}{\gamma}\right)\frac{dV}{V} - \left[\frac{2\gamma P_1 V_1 + (\gamma - 1)G^2 V_1^2}{\gamma G^2}\right]\frac{dV}{V^3} + 8j_f\left(\frac{dL}{d_i}\right) = 0$$

$$(6.6\text{--}22)$$

Integrate equation (6.6–22) for steady flow of an ideal gas flowing adiabatically between points 1 and 2 distance L apart in a horizontal pipe of constant circular cross-section to give

$$\left(\frac{\gamma + 1}{\gamma}\right)\ln\frac{V_2}{V_1} + \left[\left(\frac{\gamma - 1}{2\gamma}\right) + \frac{P_1}{V_1 G^2}\right]\left[1 - \left(\frac{V_1}{V_2}\right)^2\right] = 8j_f\left(\frac{L}{d_i}\right)$$

$$(6.6\text{--}23)$$

In terms of densities, equation (6.6–23) can be written as

$$\left(\frac{\gamma+1}{\gamma}\right)\ln\frac{\rho_1}{\rho_2}+\left[\left(\frac{\gamma-1}{2\gamma}\right)+\frac{P_1\rho_1}{G^2}\right]\left[1-\left(\frac{\rho_2}{\rho_1}\right)^2\right]=8j_f\left(\frac{L}{d_i}\right)$$

(6.6–24)

Equation (6.6–24) can be used to calculate ρ_2 knowing G and P_1. Alternatively it can be used to calculate G knowing ρ_1 and ρ_2.

The flow rate of an ideal gas for a given pressure drop is greater for adiabatic conditions than for isothermal conditions. Although the maximum possible difference is 20 per cent, for ratios $L/d_i > 1000$ the difference is seldom more than 5 per cent.

6.7 Adiabatic flow of an ideal gas through a constriction in a horizontal conduit

It has already been shown that equation (6.6–12) holds for the flow of an ideal gas through a conduit of circular cross-section under adiabatic conditions.

$$\left(\frac{\gamma}{\gamma-1}\right)(P_2V_2-P_1V_1)+\frac{(u_2^2-u_1^2)}{2}=0 \qquad (6.6–12)$$

Equation (6.2–12) holds for the reversible adiabatic change of state of an ideal gas.

$$PV^\gamma = \text{constant} \qquad (6.2–12)$$

Equation (6.2–12) can also be written in the following forms:

$$P_1V_1^\gamma = P_2V_2^\gamma \qquad (6.7–1)$$

$$\frac{V_1}{V_2}=\left(\frac{P_2}{P_1}\right)^{1/\gamma} \qquad (6.7–2)$$

$$\frac{P_2V_2}{P_1V_1}=\left(\frac{V_1}{V_2}\right)^{\gamma-1} \qquad (6.7–3)$$

$$\frac{P_2V_2}{P_1V_1}=\left(\frac{P_2}{P_1}\right)^{(\gamma-1)/\gamma} \qquad (6.7–4)$$

Substitute equation (6.7–4) into equation (6.6–12) and rearrange in the form

$$u_2^2-u_1^2=\frac{2\gamma P_1V_1}{(\gamma-1)}\left[1-\left(\frac{P_2}{P_1}\right)^{(\gamma-1)/\gamma}\right] \qquad (6.7–5)$$

The flow rate M in mass per unit time remains constant for a gas flowing steadily through a conduit in which the cross-sectional flow area S varies. For any two points 1 and 2 the following equations hold:

$$M = \rho_1 S_1 u_1 = \rho_2 S_2 u_2 \qquad (6.7\text{-}6)$$

$$M = \frac{S_1 u_1}{V_1} = \frac{S_2 u_2}{V_2} \qquad (6.7\text{-}7)$$

Rewrite equation (6.7–7) in the form

$$\frac{u_1}{u_2} = \left(\frac{S_2}{S_1}\right)\left(\frac{V_1}{V_2}\right) \qquad (6.7\text{-}8)$$

and combine with equation (6.7–2) to give

$$\frac{u_1}{u_2} = \left(\frac{S_2}{S_1}\right)\left(\frac{P_2}{P_1}\right)^{1/\gamma} \qquad (6.7\text{-}9)$$

Substitute equation (6.7–9) in equation (6.7–5) to give

$$u_2^2\left[1 - \left(\frac{S_2}{S_1}\right)^2\left(\frac{P_2}{P_1}\right)^{2/\gamma}\right] = \frac{2\gamma P_1 V_1}{(\gamma - 1)}\left[1 - \left(\frac{P_2}{P_1}\right)^{(\gamma - 1)/\gamma}\right] \qquad (6.7\text{-}10)$$

which can be written in the form

$$u_2 = \sqrt{\frac{2\gamma P_1 V_1[1 - (P_2/P_1)^{(\gamma - 1)/\gamma}]}{(\gamma - 1)[1 - (S_2/S_1)^2(P_2/P_1)^{2/\gamma}]}} \qquad (6.7\text{-}11)$$

Combine equations (6.7–6) and (6.7–11) to give

$$M = \rho_2 S_2\sqrt{\frac{2\gamma P_1 V_1[1 - (P_2/P_1)^{(\gamma - 1)/\gamma}]}{(\gamma - 1)[1 - (S_2/S_1)^2(P_2/P_1)^{2/\gamma}]}} \qquad (6.7\text{-}12)$$

where M is the gas flow rate in mass per unit time.

Equation (6.7–1) for the reversible adiabatic change of state of an ideal gas can also be written as

$$\frac{V_1}{V_2} = \frac{\rho_2}{\rho_1} = \left(\frac{P_2}{P_1}\right)^{1/\gamma} \qquad (6.7\text{-}13)$$

Combine equations (6.7–12) and (6.7–13) and write

$$M = S_2\sqrt{\frac{2\gamma P_1 \rho_1(P_2/P_1)^{2/\gamma}[1 - (P_2/P_1)^{(\gamma - 1)/\gamma}]}{(\gamma - 1)[1 - (S_2/S_1)^2(P_2/P_1)^{2/\gamma}]}} \qquad (6.7\text{-}14)$$

Equation (6.7–14) is commonly written in the form

$$M = S_2 Y_1 \sqrt{\frac{2P_1 \rho_1 [1 - (P_2/P_1)]}{[1 - (S_2/S_1)^2]}} \qquad (6.7\text{–}15)$$

where Y_1 is an expansion factor based on the upstream density ρ_1 and defined as

$$Y_1 = \sqrt{\left(\frac{\gamma}{\gamma - 1}\right)\left(\frac{P_2}{P_1}\right)^{2/\gamma}\left[\frac{1 - (P_2/P_1)^{(\gamma-1)/\gamma}}{1 - (P_2/P_1)}\right]\left[\frac{1 - (S_2/S_1)^2}{1 - (S_2/S_1)^2(P_2/P_1)^{2/\gamma}}\right]}$$

$$(6.7\text{–}16)$$

Since the pressure drop

$$\Delta P = P_1 - P_2$$
$$P_2 = P_1 - \Delta P \qquad (6.7\text{–}17)$$

and

$$\left(\frac{P_2}{P_1}\right)^{(\gamma-1)/\gamma} = \left(1 - \frac{\Delta P}{P_1}\right)^{(\gamma-1)/\gamma} \qquad (6.7\text{–}18)$$

Expand equation (6.7–18) by the binomial theorem to give

$$\left(1 - \frac{\Delta P}{P_1}\right)^{(\gamma-1)/\gamma} = 1 - \left(\frac{\gamma - 1}{\gamma}\right)\frac{\Delta P}{P_1} + \left(\frac{\gamma - 1}{2\gamma}\right)\left(-\frac{1}{\gamma}\right)\left(\frac{\Delta P}{P_1}\right)^2 + \ldots$$

$$(6.7\text{–}19)$$

When the pressure drop ΔP is small equations (6.7–18) and (6.7–19) reduce to

$$\left(\frac{P_2}{P_1}\right)^{(\gamma-1)/\gamma} = 1 - \left(\frac{\gamma - 1}{\gamma}\right)\frac{\Delta P}{P_1} \qquad (6.7\text{–}20)$$

which can be rewritten as

$$1 - \left(\frac{P_2}{P_1}\right)^{(\gamma-1)/\gamma} = \left(\frac{\gamma - 1}{\gamma}\right)\frac{\Delta P}{P_1} \qquad (6.7\text{–}21)$$

Combine equations (6.7–12) and (6.7–21) to give

$$M = \rho_2 S_2 \sqrt{\frac{2(P_1 - P_2)}{\rho_1 [1 - (S_2/S_1)^2(P_2/P_1)^{2/\gamma}]}} \qquad (6.7\text{–}22)$$

for small pressure drops.

For incompressible fluids of density ρ, equation (6.7–22) can be written in terms of the volumetric flow rate Q as

$$Q = S_2 \sqrt{\frac{2(P_1 - P_2)}{\rho[1 - (S_2/S_1)^2]}} \qquad (6.7\text{–}23)$$

When a gas discharges from a vessel through a conduit, the mean linear approach velocity u_1 is so small that it can be neglected in equation (6.7–5) which for this case can be written as

$$u_2 = \sqrt{\frac{2\gamma P_1 V_1}{(\gamma - 1)}\left[1 - \left(\frac{P_2}{P_1}\right)^{(\gamma - 1)/\gamma}\right]} \qquad (6.7\text{–}24)$$

Combine equations (6.7–6), (6.7–13) and (6.7–24) to give

$$M^2 = \left[S_2^2\left(\frac{2\gamma}{\gamma - 1}\right)P_1\rho_1\right]\left[\left(\frac{P_2}{P_1}\right)^{(2/\gamma)} - \left(\frac{P_2}{P_1}\right)^{(\gamma + 1)/\gamma}\right] \qquad (6.7\text{–}25)$$

or

$$M^2 = \left[S_2^2\left(\frac{2\gamma}{\gamma - 1}\right)P_1\rho_1\right](r^{2/\gamma} - r^{(\gamma + 1)/\gamma}) \qquad (6.7\text{–}26)$$

where $r = P_2/P_1$.

Differentiate equation (6.7–26) with respect to the pressure ratio r to give

$$2M\frac{dM}{dr} = \left[S_2^2\left(\frac{2\gamma}{\gamma - 1}\right)P_1\rho_1\right]\left[\frac{2}{\gamma}r^{[(2/\gamma) - 1]} - \left(\frac{\gamma + 1}{\gamma}\right)r^{1/\gamma}\right] \qquad (6.7\text{–}27)$$

At the maximum flow rate $dM/dr = 0$ and

$$r_c^{(\gamma - 1)/\gamma} = \frac{2}{\gamma + 1} \qquad (6.7\text{–}28)$$

where r_c is known as the critical pressure ratio P_2/P_1. Substitute equation (6.7–28) into equation (6.7–24) to give

$$u_2^2(\text{max}) = \frac{2\gamma P_1 V_1}{\gamma + 1} \qquad (6.7\text{–}29)$$

or

$$u_2^2(\text{max}) = \gamma P_1 V_1 r_c^{(\gamma - 1)/\gamma} \qquad (6.7\text{–}30)$$

Since from equation (6.7–4)

$$\frac{P_2 V_2}{P_1 V_1} = \left(\frac{P_2}{P_1}\right)^{(\gamma - 1)/\gamma} \qquad (6.7–4)$$

$$u_2(\text{max}) = \sqrt{\gamma P_2 V_2} \qquad (6.7–31)$$

Equation (6.7–31) is equivalent to equation (6.3–9)

$$u_s = \sqrt{\gamma P V} \qquad (6.3–9)$$

It has already been shown that equation (6.3–9) gives the sonic velocity for adiabatic conditions. Thus at the critical pressure ratio r_c, the mean linear velocity u_2 in the throat or constriction is the sonic velocity. This is the maximum possible velocity in a contracting nozzle, although higher gas velocities are possible in expanding nozzles.

Example (6.7–1)
Nitrogen contained in a large tank at a pressure of 200 000 N/m² and a temperature of 300 K flows steadily under adiabatic conditions into a second tank through a horizontal converging nozzle with a 0.05 m diameter throat. The pressure in the second tank and at the nozzle throat is 140 000 N/m². Calculate the flow rate of nitrogen in kg/s assuming frictionless flow and ideal gas behaviour. Also calculate the mean linear velocity in the nozzle throat and establish that this is below the critical velocity. The following data are given:

gas constant R_G = 8.3143 kJ/(kmol K)
molecular weight of nitrogen (MW) = 28.02 kg/kmol
ratio of heat capacities per unit mass for nitrogen γ = 1.39

Calculations:
flow rate

$$M = S_2 \sqrt{\frac{2\gamma P_1 \rho_1 (P_2/P_1)^{2/\gamma}[1 - (P_2/P_1)^{(\gamma - 1)/\gamma}]}{(\gamma - 1)[1 - (S_2/S_1)^2 (P_2/P_1)^{2/\gamma}]}} \qquad (6.7–14)$$

throat diameter d_2 = 0.05 m

cross-sectional flow area in throat

$$S_2 = \frac{\pi d_2^2}{4} = 1.964 \times 10^{-3} \, \text{m}^2$$

$$\gamma = 1.39$$

$$\gamma - 1 = 0.39$$

$$\frac{\gamma - 1}{\gamma} = 0.2806$$

$$\frac{1}{\gamma} = 0.7194$$

$$P_1 = 200\,000 \, \text{N/m}^2$$

$$P_2 = 140\,000 \, \text{N/m}^2$$

$$\frac{P_2}{P_1} = 0.7000$$

$$\left(\frac{P_2}{P_1}\right)^{2/\gamma} = 0.7000^{1.439} = 0.5985$$

$$\left(\frac{P_2}{P_1}\right)^{(\gamma-1)/\gamma} = 0.7000^{0.281} = 0.9046$$

$$1 - \left(\frac{P_2}{P_1}\right)^{(\gamma-1)/\gamma} = 0.0954$$

Since cross-section of tank, S_1, is large assume $S_2/S_1 \cong 0$; calculate the density ρ_1 from equation (6.2–4)

$$PV = \frac{R_G T}{(MW)} \qquad (6.2\text{–}4)$$

rewrite equation (6.2–4) in the form

$$\rho_1 = \frac{P_1(MW)}{R_G T_1}$$

$$P_1 = 200\,000 \, \text{N/m}^2$$

$$(MW) = 28.02 \, \text{kg/kmol}$$

$$R_G = 8.3143 \, \text{kJ/(kmol K)}$$

$$T_1 = 300 \, \text{K}$$

$$\rho_1 = 2.247 \, \text{kg/m}^3$$

flow rate

$$M = (1.964 \times 10^{-3} \, \text{m}^2)$$

$$\times \sqrt{\frac{(2)(1.39)(200\,000 \, \text{N/m}^2)(2.247 \, \text{kg/m}^3)(0.5985)(0.0954)}{0.39}}$$

$$= (1.964 \times 10^{-3} \, \text{m}^2)\sqrt{1.829 \times 10^5 \, \text{kg}^2/(\text{s}^2 \, \text{m}^4)}$$

$$= 0.8399 \, \text{kg/s}$$

calculate the temperature T_2 of the nitrogen in the nozzle throat from equation (6.2–13)

$$\frac{P_2}{P_1} = \left(\frac{T_2}{T_1}\right)^{\gamma/(\gamma-1)} \tag{6.2–13}$$

rewrite equation (6.2–13) in the form

$$T_2 = T_1\left(\frac{P_2}{P_1}\right)^{(\gamma-1)/\gamma}$$

$$T_2 = (300 \, \text{K})(0.9046)$$

$$= 271.4 \, \text{K}$$

calculate the density ρ_2 from equation (6.2–4)

$$PV = \frac{R_G T}{(MW)} \tag{6.2–4}$$

rewrite equation (6.2–4) in the form

$$\rho_2 = \frac{P_2(MW)}{R_G T_2}$$

$$P_2 = 140\,000 \, \text{N/m}^2$$

$$(MW) = 28.02 \, \text{kg/kmol}$$

$$R_G = 8.3143 \, \text{kJ/(kmol K)}$$

$$T_2 = 271.4 \, \text{K}$$

$$\rho_2 = 1.738 \, \text{kg/m}^3$$

calculate the linear velocity u_2 in the nozzle throat from equation (6.7–6)

$$M = \rho_1 S_1 u_1 = \rho_2 S_2 u_2 \qquad (6.7\text{–}6)$$

$$u_2 = \frac{M}{\rho_2 S_2}$$

$$u_2 = \frac{0.8399 \text{ kg/s}}{(1.738\,\text{kg/m}^3)(1.964 \times 10^{-3}\,\text{m}^2)}$$

$$= 246.1 \text{ m/s}$$

calculate the maximum possible linear velocity in the nozzle throat from equation (6.7–31)

$$u_2(\text{max}) = \sqrt{\gamma P_2 V_2} \qquad (6.7\text{–}31)$$

$$u_2(\text{max}) = \sqrt{\gamma \frac{P_2}{\rho_2}}$$

$$= \sqrt{\frac{(1.39)(140\,000 \text{ N/m}^2)}{1.738 \text{ kg/m}^3}}$$

$$= 334.6 \text{ m/s}$$

thus actual $u_2 < u_2(\text{max})$.

6.8   Gas compression and compressors

Compressors are devices for supplying energy or pressure head to a gas. For the most part, compressors like pumps can be classified into centrifugal and positive displacement compressors. Centrifugal compressors impart a high velocity to the gas and the resultant kinetic energy provides the work for compression. Positive displacement compressors include rotary and reciprocating compressors although the latter are the most important for high pressures.

From equation (6.1–11) the shaft work of compression ΔW per unit mass of gas assuming a reversible frictionless process is

$$\Delta W = \int_1^2 V \, dP \qquad (6.8\text{–}1)$$

Although isothermal compression is desirable, in practice the heat of compression is never removed fast enough to make it possible.

In actual compressors only a small fraction of the heat of compression is removed and the process is almost adiabatic.

When ideal gases are compressed under reversible adiabatic conditions, they obey equation (6.2–12) which can be written in the forms

$$P_1 V_1^\gamma = P V^\gamma \tag{6.8–2}$$

and

$$\frac{P_1^{1/\gamma} V_1}{P^{1/\gamma}} = V \tag{6.8–3}$$

Substitute equation (6.8–3) into equation (6.8–1) and integrate to give

$$\Delta W = \left(\frac{\gamma}{\gamma-1}\right) P_1 V_1 \left[\left(\frac{P_2}{P_1}\right)^{(\gamma-1)/\gamma} - 1\right] \tag{6.8–4}$$

Equation (6.8–4) gives the theoretical adiabatic work of compression from pressures P_1 to P_2.

For two stage compression from pressures P_1 to P_2 to P_3, equation (6.8–4) becomes

$$\Delta W = \left(\frac{\gamma}{\gamma-1}\right) P_1 V_1 \left\{\left[\left(\frac{P_2}{P_1}\right)^{(\gamma-1)/\gamma} - 1\right] + \left[\left(\frac{P_3}{P_2}\right)^{(\gamma-1)/\gamma} - 1\right]\right\} \tag{6.8–5}$$

For n stage compression from pressures P_1 to P_2, equation (6.8–4) becomes for equal pressure ratios in each stage

$$\Delta W = \left(\frac{n\gamma}{\gamma-1}\right) P_1 V_1 \left[\left(\frac{P_2}{P_1}\right)^{(\gamma-1)/n\gamma} - 1\right] \tag{6.8–6}$$

In practice it is possible to approach more isothermal conditions by carrying out a compression in a number of stages.

When ideal gases are compressed under reversible adiabatic conditions the temperature increase from T_1 to T_2 is given by equation (6.2–13).

$$\frac{P_2}{P_1} = \left(\frac{T_2}{T_1}\right)^{\gamma/(\gamma-1)} \tag{6.2–13}$$

So far only reversible adiabatic compression of an ideal gas has been considered. For the irreversible adiabatic compression of an actual gas, the shaft work ΔW per unit mass of gas can be obtained from equation (6.1–6) which for this case becomes

$$\Delta W = H_2 - H_1 \tag{6.8-7}$$

where H is the enthalpy per unit mass of gas.

The actual work of compression is greater than the theoretical work because of clearance gases, back leakage and frictional effects.

Example (6.8–1)

Calculate the theoretical work in J/kg required to compress a diatomic ideal gas initially at a temperature of 200 K adiabatically from a pressure of 10 000 N/m² to a pressure of 100 000 N/m² in (i) a single stage compressor, (ii) a compressor with two equal stages and (iii) a compressor with three equal stages. The following data are given:

gas constant R_G = 8.3143 kJ/(kmol K)
molecular weight of ideal gas (MW) = 28.00 kg/kmol
ratio of heat-capacities per unit mass for diatomic gas γ = 1.40

Calculations:
work in single stage adiabatic compression for an ideal gas

$$\Delta W = \left(\frac{\gamma}{\gamma - 1}\right) P_1 V_1 \left[\left(\frac{P_2}{P_1}\right)^{(\gamma - 1)/\gamma} - 1\right] \tag{6.8-4}$$

$$\gamma = 1.40$$

$$\gamma - 1 = 0.40$$

$$\frac{\gamma}{\gamma - 1} = 3.5$$

$$\frac{\gamma - 1}{\gamma} = 0.2857$$

$$P_1 = 10\,000 \text{ N/m}^2$$

$$P_2 = 100\,000 \text{ N/m}^2$$

$$\frac{P_2}{P_1} = 10$$

$$\left(\frac{P_2}{P_1}\right)^{(\gamma - 1)/\gamma} = 10^{0.2857} = 1.931$$

$$PV = \frac{R_G T}{(MW)} \qquad (6.2\text{--}4)$$

$$P_1 V_1 = \frac{R_G T}{(MW)}$$

$$P_1 V_1 = \frac{[8.3143 \text{ kJ/(kmol K)}](200 \text{ K})}{28.00 \text{ kg/kmol}}$$

$$= 59.34 \text{ kJ/kg}$$

$$\Delta W \text{ (1 stage)} = (3.5)(59.34 \text{ kJ/kg})(1.931 - 1)$$

(i) ΔW (1 stage) = 193.4 kJ/kg

work in n stage adiabatic compression for an ideal gas

$$\Delta W = \left(\frac{n\gamma}{\gamma - 1}\right) P_1 V_1 \left[\left(\frac{P_2}{P_1}\right)^{(\gamma - 1)/n\gamma} - 1\right] \qquad (6.8\text{--}6)$$

for $n = 2$

$$\frac{n\gamma}{\gamma - 1} = 7.0$$

$$\frac{\gamma - 1}{n\gamma} = 0.1428$$

$$\left(\frac{P_2}{P_1}\right)^{(\gamma - 1)/n\gamma} = 10^{0.1428} = 1.390$$

$$P_1 V_1 = 59.34 \text{ kJ/kg}$$

$$\Delta W \text{ (2 stages)} = (7.0)(59.34 \text{ kJ/kg})(1.390 - 1)$$

(ii) ΔW (2 stages) = 162.0 kJ/kg

for $n = 3$

$$\frac{n\gamma}{\gamma - 1} = 10.5$$

$$\frac{\gamma - 1}{n\gamma} = 0.0953$$

$$\left(\frac{P_2}{P_1}\right)^{(\gamma-1)/n\gamma} = 10^{0.0952} = 1.246$$

$$P_1 V_1 = 59.34 \text{ kJ/kg}$$

$$\Delta W \text{ (3 stages)} = (10.5)(59.34 \text{ kJ/kg})(1.246 - 1)$$

(iii) ΔW (3 stages) = 153.3 kJ/kg

REFERENCES

(1) Barna, P. S., *Fluid Mechanics for Engineers*, 3rd ed., p. 248, London, Butterworths, 1969.
(2) Dodge, B. F., *Chemical Engineering Thermodynamics*, p. 100, New York, McGraw-Hill Book Co. Inc., 1944.
(3) Lapple, C. E., Trans. Am. Inst. Chem. Eng., **39**, 385 (1943).
(4) Perry, J. H., *Chemical Engineers' Handbook*, 4th ed., p. 4–50, New York, McGraw-Hill Book Co. Inc., 1963.
(5) Shapiro, A. H., *The Dynamics and Thermodynamics of Compressible Fluid Flow*, p. 47, New York, The Ronald Press Company, 1953.

7
Flow of two phase gas liquid mixtures in pipes

7.1 Flow patterns for two phase gas liquid flow

When a gas and a liquid flow cocurrently in a pipe, both the gas and the liquid may be in turbulent flow or they may both be in laminar flow. Alternatively the gas may be in turbulent flow and the liquid in laminar flow or vice versa. Flow patterns for gas liquid systems have been extensively observed for both horizontal and vertical flow.[1,6]

As the flow rate is increased in a horizontal pipe, the flow passes through the following stages[1]: bubble flow where the bubbles flow along the upper surface of the pipe, plug flow where the gas bubbles coalesce into plugs, stratified flow where the gas liquid interface is smooth and well defined, wave flow where surface waves are formed, slug flow where the waves touch the top of the pipe and form a frothy slug, annular flow where the gas flows as a core surrounded by a wall of liquid, and mist flow where the gas core entrains the liquid from the pipe wall.

As the flow rate is increased in a vertical pipe, the flow passes through the following stages[6]: bubble flow in which the gas exists as dispersed bubbles, slug flow where alternate slugs of gas and liquid move up the pipe, froth flow where the gas bubbles intermix with the liquid, annular flow where the gas flows as a core surrounded by a wall of more slowly moving liquid, and mist flow where the gas core entrains the liquid from the pipe wall.

The availability of this detailed visual information on flow has not formed the basis of a theoretical method for the prediction of pressure drops in gas liquid flow systems. A number of empirical

133

prediction methods have been developed.[2-5] Most of these ignore the diversity of flow regimes.

The pressure drop in a pipe in which gas and liquid flow cocurrently is always greater than when either of the two phases flows by itself at the same rate. The reason for this is that energy is required for the formation of the gas-liquid interface.

7.2   Prediction of pressure drop by the Lockhart and Martinelli method when both phases are turbulent[5]

Consider a pipe in which the gas, liquid, and total flow rates in mass per unit time are M_G, M_L and M_T respectively.

$$M_T = M_G + M_L \qquad (7.2\text{--}1)$$

Let the inside diameter and cross-sectional flow area in the pipe be d_i and S_i respectively.

Let the weight fraction of gas or vapour be w_G. This fraction is also commonly referred to as quality.

$$M_G = M_T w_G \qquad (7.2\text{--}2)$$

$$M_L = M_T(1 - w_G) \qquad (7.2\text{--}3)$$

The mean linear velocities for the gas and liquid in the pipe are given respectively by the following equations:

$$u_G = \frac{M_T w_G}{S_i \rho_G} \qquad (7.2\text{--}4)$$

$$u_L = \frac{M_T(1 - w_G)}{S_i \rho_L} \qquad (7.2\text{--}5)$$

The pressure drop in a pipe for single phase flow is given by equation (2.4–3)

$$\Delta P = 8j_f \left(\frac{L}{d_i}\right)\frac{\rho u^2}{2} \qquad (2.4\text{--}3)$$

In this case, for the flow of the liquid phase only, equation (2.4–3) can be written in the form

$$\left(\frac{\Delta P}{L}\right)_f = 8(j_f)_L \left(\frac{1}{d_i}\right)\frac{\rho_L u_L^2}{2} \qquad (7.2\text{--}6)$$

Substitute equation (7.2–5) into equation (7.2–6) to give

$$\left(\frac{\Delta P}{L}\right)_f = 8(j_f)_L \left(\frac{1}{d_i}\right) \left[\left(\frac{M_T}{S_i}\right)^2 \left(\frac{1}{2\rho_L}\right)\right] (1 - w_G)^2 \qquad (7.2\text{–}7)$$

Equation (7.2–7) gives the pressure drop per unit length of pipe due to friction for the liquid flowing by itself in the pipe at the same rate. The actual pressure drop for gas liquid flow is greater than this by

TABLE (7.2–1)

X_{tt}	ϕ
0.01	128
0.02	68.4
0.04	38.5
0.07	24.4
0.10	18.5
0.2	11.2
0.4	7.05
0.7	5.04
1.0	4.20
2.0	3.10
4.0	2.38
7.0	1.96
10	1.75
20	1.48
40	1.29
70	1.17
100	1.11

a factor ϕ^2. Values of ϕ are given in Table (7.2–1)[5] for various values of X_{tt} where

$$X_{tt} = \left[\frac{(1 - w_G)}{w_G}\right]^{0.9} \left(\frac{\rho_G}{\rho_L}\right)^{0.5} \left(\frac{\mu_L}{\mu_G}\right)^{0.1} \qquad (7.2\text{–}8)$$

Thus the pressure drop per unit length of pipe due to friction for gas liquid cocurrent flow is given by the equation

$$\left(\frac{\Delta P}{L}\right)_f = \phi^2 8(j_f)_L \left(\frac{1}{d_i}\right) \left[\left(\frac{M_T}{S_i}\right)^2 \left(\frac{1}{2\rho_L}\right)\right] (1 - w_G)^2 \qquad (7.2\text{–}9)$$

The Reynolds number for the liquid flow is

$$(N_{RE})_L = \frac{\rho_L u_L d_i}{\mu_L} \qquad (7.2\text{–}10)$$

Substitute equation (7.2–5) into equation (7.2–10) and write

$$(N_{RE})_L = \frac{M_T d_i (1 - w_G)}{S_i \mu_L} \qquad (7.2\text{–}11)$$

The friction factor $(j_f)_L$ in equation (7.2–9) can be read from Figure (2.4–1) for the value of the Reynolds number $(N_{RE})_L$ given by equation (7.2–11) knowing the dimensionless roughness factor for the pipe ε/d_i.

The Lockhart–Martinelli correlation is based on work done in horizontal pipes varying in internal diameter from 0.0013 m to 0.026 m.

Example (7.2–1)

A mixture of gas and liquid flows through a tube of internal diameter 0.02 m at a steady total flow rate of 0.2 kg/s. The tube roughness $\varepsilon = 0.0000015$ m. The dynamic viscosities of the gas and liquid are 1.0×10^{-5} N s/m^2 and 2.0×10^{-3} N s/m^2 respectively. The densities of the gas and liquid are 60 kg/m^3 and 1000 kg/m^3 respectively. The weight fraction of gas is 0.149. Calculate the pressure gradient in the pipe using the Lockhart–Martinelli correlation.

Calculations:

$$\text{Reynolds number } (N_{RE})_L = \frac{M_T d_i (1 - w_G)}{S_i \mu_L} \qquad (7.2\text{–}11)$$

$$M_T = 0.2 \text{ kg/s}$$

$$d_i = 0.02 \text{ m}$$

$$1 - w_G = 0.851$$

$$S_i = 0.0003142 \text{ m}^2$$

$$\mu_L = 0.0020 \text{ N s/m}^2$$

$$(N_{RE})_L = \frac{(0.2 \text{ kg/s})(0.02 \text{ m})(0.851)}{(0.0003142 \text{ m}^2)(0.0020 \text{ N s/m}^2)}$$

$$= 5417$$

from j_f against N_{RE} graph in Figure (2.4–1), $(j_f)_L = 0.0043$ for $N_{RE} = 5417$ and $\varepsilon/d_i = 0.0000015$ m/0.02 m = 0.000075.

Lockhart–Martinelli parameter

$$X_{tt} = \left[\frac{(1 - w_G)}{w_G}\right]^{0.9}\left(\frac{\rho_G}{\rho_L}\right)^{0.5}\left(\frac{\mu_L}{\mu_G}\right)^{0.1} \qquad (7.2-8)$$

$$\frac{(1 - w_G)}{w_G} = 5.711$$

$$\left[\frac{(1 - w_G)}{w_G}\right]^{0.9} = 4.797$$

$$\rho_G = 60 \text{ kg/m}^3$$

$$\rho_L = 1000 \text{ kg/m}^3$$

$$\frac{\rho_G}{\rho_L} = \frac{6}{100}$$

$$\left(\frac{\rho_G}{\rho_L}\right)^{0.5} = \frac{2.449}{10} = 0.2449$$

$$\mu_L = 2.0 \times 10^{-3} \text{ N s/m}^2$$

$$\mu_G = 1.0 \times 10^{-5} \text{ N s/m}^2$$

$$\frac{\mu_L}{\mu_G} = 200$$

$$\left(\frac{\mu_L}{\mu_G}\right)^{0.1} = 1.698$$

$$X_{tt} = (4.797)(0.2449)(1.698)$$

$$= 1.995$$

from Table (7.2–1)

$$\phi = 3.106 \text{ for } X_{tt} = 1.995$$

$$\phi^2 = 9.647$$

pressure gradient

$$\left(\frac{\Delta P}{L}\right)_f = \phi^2 8(j_f)_L\left(\frac{1}{d_i}\right)\left[\left[\left(\frac{M_T}{S_i}\right)^2\left(\frac{1}{2\rho_L}\right)\right]\right](1 - w_G)^2 \qquad (7.2-9)$$

$$\phi^2 = 9.647$$

$$(j_f)_L = 0.0043$$

$$d_i = 0.02 \text{ m}$$

$$\frac{M_T}{S_i} = \frac{0.2 \text{ kg/s}}{0.0003142 \text{ m}^2} = 636.5 \text{ kg/(s m}^2)$$

$$\rho_L = 1000 \text{ kg/m}^3$$

$$(1 - w_G) = 0.851$$

$$\left(\frac{\Delta P}{L}\right)_f = (9.647)(8)(0.0043)\left(\frac{1}{0.02 \text{ m}}\right)$$

$$\times \left(636.5\frac{\text{kg}}{\text{s m}^2}\right)^2 \left(\frac{1}{2000 \text{ kg/m}^3}\right)(0.851)^2$$

$$= 2434 \text{ (N/m}^2)/\text{m}$$

REFERENCES

(1) Alves, G. E., Chem. Eng. Prog., **50**, 449 (1954).
(2) Chenoweth, J. M., and Martin, M. W., Pet. Refiner, **34**, No. 10, 151 (1955).
(3) Dukler, A. E., Wicks, M., and Cleveland, R. G., AIChEJ., **10**, No. 1, 38 (1964).
(4) Dukler, A. E., Wicks, M., and Cleveland, R. G., AIChEJ., **10**, No. 1, 44 (1964).
(5) Lockhart, R. W., and Martinelli, R. C., Chem. Eng. Prog., **45**, 39 (1949).
(6) Nicklin, D. J., and Davidson, J. F., *Symposium on Two-Phase Flow*, Paper No. 4, Inst. Mech. Eng., London, Feb. 1962.

8
Flow measurement

8.1 Flowmeters and flow measurement

The flow of fluids is most commonly measured using head flowmeters. The operation of these flowmeters is based on the Bernoulli equation. A constriction in the flow path is used to increase the linear flow velocity. This is accompanied by a decrease in pressure head and since the resultant pressure drop is a function of the flow rate of fluid, the latter can be evaluated. The flowmeters for closed conduits can be used for both gases and liquids. The flowmeters for open conduits can only be used for liquids. Head flowmeters include orifice and venturi meters, flow nozzles, Pitot tubes and weirs. They consist of a primary element which causes the pressure or head loss and a secondary element which measures it. The primary element does not contain any moving parts. The most common secondary element for closed conduit flowmeters is a U-tube manometer.

A U-tube manometer is shown schematically in Figure (8.1–1). One arm is connected to the high pressure tap and the other arm to the low pressure tap in the flowing fluid. The fluid in one arm of the manometer is separated from the fluid in the other arm by an immiscible liquid of higher density which is usually mercury. Consider the pressures at levels a and b in the two arms of the manometer shown in Figure (8.1–1) when the system is in equilibrium. Let the pressures at level b be P_1 and P_2 in arm 1 and arm 2, respectively. Let the difference in the heights of immiscible liquid in the two arms of the manometer be Δz.

139

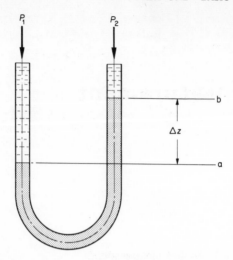

Figure (8.1–1)
U-tube manometer.

The pressure at level a in the manometer is $(P_1 + \rho \, \Delta z \, g)$ in arm 1 and $(P_2 + \rho_m \, \Delta z \, g)$ in arm 2 where ρ and ρ_m are the densities of the flowing fluid and immiscible liquid respectively. These two pressures are equal since the two arms of the manometer are connected by a continuous column of liquid.

Therefore

$$P_1 + \rho \, \Delta z \, g = P_2 + \rho_m \, \Delta z \, g \qquad (8.1\text{–}1)$$

which can be written as

$$P_1 - P_2 = (\rho_m - \rho) \, \Delta z \, g \qquad (8.1\text{–}2)$$

If ρ and ρ_m are in kg/m^3, Δz is in m, and g is 9.81 m/s^2, the pressure differential across the primary element $P_1 - P_2$ is in N/m^2. The head differential across the primary element Δh based on the flowing fluid is

$$\Delta h = \frac{P_1 - P_2}{\rho g} \qquad (8.1\text{–}3)$$

Combine equations (8.1–2) and (8.1–3) to give

$$\Delta h = \frac{(\rho_m - \rho) \, \Delta z}{\rho} \qquad (8.1\text{–}4)$$

which relates the head differential across the primary element to the difference in height of immiscible liquid in the two arms of the manometer.

Other flowmeters are in common use which operate on principles differing from head flowmeters. Mechanical flowmeters have primary elements which contain moving parts. These flowmeters include rotameters, positive displacement meters and velocity meters. Electromagnetic flowmeters have the advantages of no restriction in a conduit and no moving parts.

8.2 Head flowmeters in closed conduits

The primary element of an orifice meter is simply a flat plate containing a drilled hole located in a pipe perpendicular to the direction of fluid flow as shown in Figure (8.2–1).

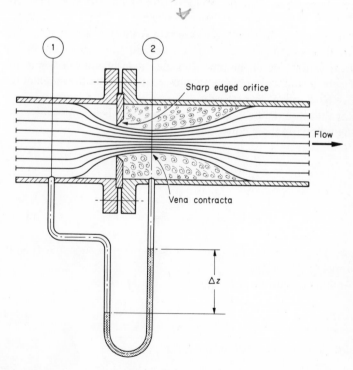

Figure (8.2–1)
Orifice meter.

Equation (1.6–7) is the modified Bernoulli equation for steady flow in a pipe.

$$\left(z_2 + \frac{P_2}{\rho_2 g} + \frac{u_2^2}{2g\alpha_2}\right) - \left(z_1 + \frac{P_1}{\rho_1 g} + \frac{u_1^2}{2g\alpha_1}\right) = \Delta h - h_f \qquad (1.6\text{–}7)$$

For the steady horizontal flow of an incompressible fluid of density ρ between points 1 and 2 in a pipe with no pump and and no friction, equation (1.6–7) can be written as

$$\frac{u_2^2}{2\alpha_2}\left(1 - \frac{\alpha_2 u_1^2}{\alpha_1 u_2^2}\right) = \frac{(P_1 - P_2)}{\rho} \qquad (8.2\text{–}1)$$

Consider points 1 and 2 in Figure (8.2–1). At point 1 in the pipe, the fluid flow is undisturbed by the orifice plate. The fluid at this point has a mean linear velocity u_1 and a cross-sectional flow area S_1. At point 2 in the pipe the fluid attains its maximum mean linear velocity u_2 and its smallest cross-sectional flow area S_2. This point is known as the vena contracta. It occurs at about one half to two pipe diameters downstream from the orifice plate. The location is a function of the flow rate and the size of the orifice relative to the size of the pipe. Let the mean linear velocity in the orifice be u' and let the diameter and cross-sectional flow area of the orifice be d' and S' respectively.

For this case the principle of continuity can be expressed by any of the following three equations.

$$M = \rho S_1 u_1 = \rho S_2 u_2 = \rho S' u' \qquad (8.2\text{–}2)$$

$$Q = S_1 u_1 = S_2 u_2 = S' u'$$

or

$$Q = \frac{\pi}{4}d_1^2 u_1 = \frac{\pi}{4}d_2^2 u_2 = \frac{\pi}{4}d'^2 u' \qquad (8.2\text{–}3)$$

where M is the flow rate of fluid in mass per unit time and Q is the volumetric flow rate.

Combine equations (8.2–1) and (8.2–3) and write

$$\frac{u'^2}{2\alpha_2}\left(\frac{d'}{d_2}\right)^4\left[1 - \frac{\alpha_2}{\alpha_1}\left(\frac{d_2}{d_1}\right)^4\right] = \frac{(P_1 - P_2)}{\rho} \qquad (8.2\text{–}4)$$

which can be rearranged in the form

$$u' = \left(\frac{d_2}{d'}\right)^2 \sqrt{\frac{2(P_1 - P_2)\alpha_2}{\rho[1 - (\alpha_2/\alpha_1)(d_2/d_1)^4]}} \qquad (8.2\text{–}5)$$

Combine equations (8.2–3) and (8.2–5) and write

$$Q = S' \left(\frac{d_2}{d'}\right)^2 \sqrt{\frac{2(P_1 - P_2)\alpha_2}{\rho[1 - (\alpha_2/\alpha_1)(d_2/d_1)^4]}} \qquad (8.2\text{–}6)$$

Equation (8.2–6) gives the volumetric flow rate Q when there is no friction in the system. For turbulent flow, equation (8.2–6) reduces to equation (6.7–23).

In practice, the measured volumetric flow rate Q is always less than Q given by equation (8.2–6). Viscous frictional effects retard the flowing fluid. In addition, boundary layer separation occurs on the downstream side of the orifice plate resulting in a substantial pressure or head loss from form friction. This effect is a function of the geometry of the system.

In practice, the volumetric flow rate Q is given by equation (8.2–7)

$$Q = S'C_d \sqrt{\frac{2(P_1 - P_2)}{\rho[1 - (d'/d_1)^4]}} \qquad (8.2\text{–}7)$$

In equation (8.2–7) which is analogous to equation (8.2–6), C_d is the dimensionless discharge coefficient which accounts for geometry and friction; d'/d_1 is the ratio of the diameter of the orifice hole to the inside diameter of the pipe. This ratio does not vary as does the ratio d_2/d_1 in equation (8.2–6) for frictionless flow.

Substitute equation (8.1–3) into equation (8.2–7) to give

$$Q = S'C_d \sqrt{\frac{2g\,\Delta h}{[1 - (d'/d_1)^4]}} \qquad (8.2\text{–}8)$$

Equation (8.2–8) gives the volumetric flow rate Q in terms of the head differential across the orifice plate Δh. The latter is based on the flowing fluid.

Both equations (8.2–7) and (8.2–8) refer to horizontal pipes. For the case of a fluid flowing upward through an orifice meter in a pipe which is not horizontal, equation (8.2–8) must be written in the modified form

$$Q = S'C_d \sqrt{\frac{2g(\Delta h + \Delta z_{12})}{[1 - (d'/d_1)^4]}} \qquad (8.2\text{–}9)$$

where Δz_{12} is the difference between the heights of the pressure taps at points 1 and 2.

Concentric

Eccentric

Segmental

Figure (8.2–2)
Concentric, eccentric and segmental orifice plates.

The holes in orifice plates may be either concentric, eccentric or segmental as shown in Figure (8.2–2). Orifice plates are prone to damage by erosion.

The coefficient of discharge C_d for a particular orifice meter is a function of the location of the pressure taps, the ratio of the diameter of the orifice hole to the inside diameter of the pipe d'/d_1, the Reynolds number in the pipeline N_{RE}, and the thickness of the orifice plate. Authoritative references should be consulted for values of C_d.[6] Data are frequently given as log–log plots of C_d against N_{RE}. It should be noted whether the Reynolds numbers are based on the inside diameter of the pipe or the diameter of the orifice hole. The most common range for C_d is 0.6 to 0.7.

Orifice meters suffer from high frictional pressure or head losses. Thus, most of the pressure drop is not recoverable. The pressure loss is given by the equation[1]

$$\Delta P = \left[1 - \left(\frac{d'}{d_1}\right)^2\right](P_1 - P_2) \qquad (8.2\text{–}10)$$

Orifice plates are inexpensive and easy to install since they can readily be inserted at a flanged joint.

Figure (8.2–3) shows a Venturi meter. The theory is the same as for the orifice meter but a much higher proportion of the pressure drop is recoverable than is the case with orifice meters. The gradual approach to and the gradual exit from the orifice substantially eliminates boundary layer separation. Thus, form drag and eddy formation are reduced to a minimum.

A series of tap connections in an annular pressure ring gives a mean value for the pressure at point 1 in the approach section and also at point 2 in the throat. Although Venturi meters are relatively expensive and tend to be bulky, they can meter up to 60 per cent

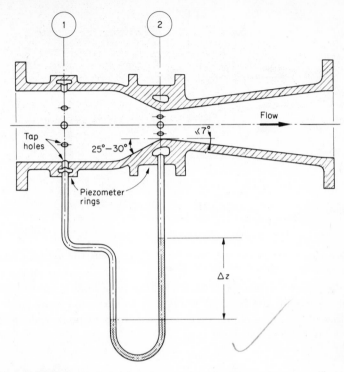

Figure (8.2–3)
Venturi meter.

more flow than orifice plates for the same inside pipe diameter and differential pressure.[4] The coefficient of discharge C_d for a Venturi meter is in the region of 0.98. Venturies are more suitable than orifice plates for metering liquids containing solids.

Figure (8.2–4) shows a flow nozzle. This is a modified and less expensive type of Venturi meter.

The theoretical treatment of head flowmeters in this section is for incompressible fluids. The flow of compressible fluids through a constriction in a pipe is treated in Chapter 6.

Orifice meters, Venturi meters and flow nozzles measure volumetric flow rate Q or mean linear velocity u. In contrast the Pitot tube shown in a horizontal pipe in Figure (8.2–5) measures a point velocity v. Thus Pitot tubes can be used to obtain velocity profiles in either open or closed conduits. At point 1 in Figure (8.2–5) a small amount of fluid is brought to a standstill. Thus the combined

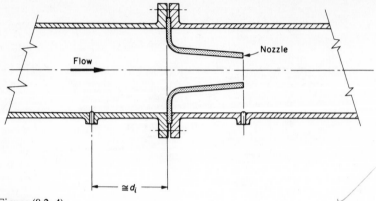

Figure (8.2–4)
Flow nozzle.

head at point 1 is the pressure head $P/(\rho g)$ plus the velocity head $v^2/(2g)$ if the potential head z at the centre of the horizontal pipe is arbitrarily taken to be zero. Since at point 2 fluid is not brought to a standstill, the head at point 2 is the pressure head only if points 1 and 2 are sufficiently close for them to be considered to have the same potential head z.

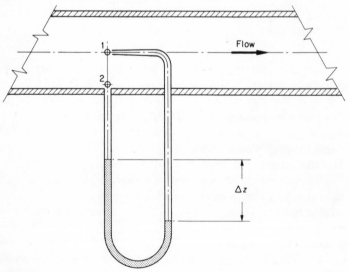

Figure (8.2–5)
Pitot tube.

Thus the difference in head $\Delta h'$ between points 1 and 2 neglecting friction is the velocity head $v^2/(2g)$. Therefore the point velocity v is given by the equation

$$v = \sqrt{2g\,\Delta h'} \qquad (8.2\text{–}11)$$

The difference in heads between points 1 and 2 $\Delta h'$ is usually measured with a manometer.

Combine equations (8.1–4) and (8.2–11) to give

$$v = \sqrt{\frac{2g(\rho_m - \rho)\,\Delta z'}{\rho}} \qquad (8.2\text{–}12)$$

Equation (8.2–12) gives the point velocity v in terms of the difference in level between the two arms of the manometer $\Delta z'$, the density of the flowing fluid ρ, the density of the immiscible manometer liquid ρ_m and the gravitational acceleration g. If ρ and ρ_m are in kg/m^3, $g = 9.81$ m/s, $\Delta z'$ is in m the point velocity v is in m/s.

Most Pitot tubes consist of two concentric tubes parallel to the direction of fluid flow. The inner tube points into the flow and the outer tube is perforated with small holes which are perpendicular to the direction of flow. The inner tube transmits the combined pressure and velocity heads and the outer tube only the pressure head.

Although Pitot tubes are inexpensive and have negligible permanent head losses they are not widely used. They are highly sensitive to fouling, their required alignment is critical and they cannot measure volumetric flow rate Q or mean linear velocity u. The latter can be calculated from a single measurement only if the velocity distribution is known.

Example (8.2–1)

Calculate the volumetric flow rate of water in m^3/s through a pipe with an inside diameter of 0.15 m fitted with an orifice plate containing a concentric hole of diameter 0.10 m given the following data:

(1) A 0.254 m difference in level on a mercury manometer connected across the orifice plate.
(2) A mercury specific gravity of 13.6.
(3) A discharge coefficient of 0.60.

Calculations:

$$\Delta h = \frac{(\rho_m - \rho)\Delta z}{\rho} \tag{8.1-4}$$

$$= \left(\frac{\rho_m}{\rho} - 1\right)\Delta z$$

$$\Delta z = 0.254 \text{ m}$$

$$\frac{\rho_m}{\rho} = 13.6$$

head differential across orifice

$$\Delta h = (13.6 - 1)(0.254 \text{ m})$$
$$= 3.20 \text{ m}$$

$$Q = S'C_d\sqrt{\frac{2g\,\Delta h}{[1 - (d'/d_1)^4]}} \tag{8.2-8}$$

$$S' = \frac{\pi d'^2}{4} = \frac{(3.142)(0.10 \text{ m})^2}{4} = 0.007855 \text{ m}^2$$

$$g = 9.81 \text{ m/s}^2$$

$$\Delta h = 3.20 \text{ m}$$

$$\frac{d'}{d_1} = \frac{0.10}{0.15} = \frac{1}{1.5}$$

$$1 - \left(\frac{d'}{d_1}\right)^4 = 0.8025$$

$$\sqrt{\frac{2g\,\Delta h}{[1 - (d'/d_1)^4]}} = \sqrt{\frac{(2)(9.81 \text{ m/s}^2)(3.20 \text{ m})}{0.8025}} = 8.845 \text{ m/s}$$

$$C_d = 0.60$$

volumetric flow rate

$$Q = (0.007855 \text{ m}^2)(0.60)(8.845 \text{ m/s})$$
$$= 0.417 \text{ m}^3/\text{s}$$

8.3   Head flowmeters in open conduits

Weirs are commonly used to measure the flow rate of liquids in open conduits. The theory is based on the Bernoulli equation for frictionless flow.

$$\left(z_2 g + \frac{P_2}{\rho_2} + \frac{v_2^2}{2}\right) - \left(z_1 g + \frac{P_1}{\rho_1} + \frac{v_1^2}{2}\right) = 0 \qquad (1.6\text{--}4)$$

Consider a liquid flowing over a sharp crested weir as shown in Figure (8.3–1). Let the upstream level of the liquid be z_o above the level of the weir crest. As the liquid approaches the weir, the liquid

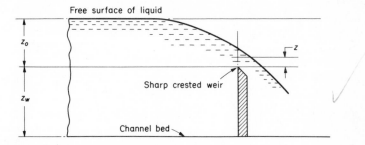

Figure (8.3–1)
Flow of liquid over a sharp crested weir.

level gradually drops and the flow velocity increases. Downstream from the weir, a jet is formed. This is called the nappe and it is ventilated underneath to enable it to spring free from the weir crest.[2]

Consider any point in the liquid at a height z vertically above the weir crest in Figure (8.3–1). Let the liquid at this point have a linear velocity v. For this case the Bernoulli equation can be written in the form

$$\left(z_o g + \frac{v_o^2}{2}\right) - \left(z g + \frac{v^2}{2}\right) = 0 \qquad (8.3\text{--}1)$$

Equation (8.3–1) is based on the following assumptions: the approach velocity v_o is uniform and parallel, the streamlines are horizontal above the weir, there is atmospheric pressure under the nappe, and the flow is frictionless.

Rewrite equation (8.3–1) in the form

$$v = \sqrt{(2g)}\sqrt{\left(z_o + \frac{v_o^2}{2g} - z\right)} \tag{8.3-2}$$

Equation (8.3–2) gives the point linear velocity v at a height z above the weir crest. If the width of the conduit is b at this point, the volumetric flow rate through an element of cross-sectional flow area of height dz is

$$dQ = bv\,dz \tag{8.3-3}$$

Substitute equation (8.3–2) into equation (8.3–3) and write

$$dQ = b\sqrt{(2g)}\sqrt{\left(z_o + \frac{v_o^2}{2g} - z\right)}\,dz \tag{8.3-4}$$

Let

$$h = z_o + \frac{v_o^2}{2g} - z \tag{8.3-5}$$

so that

$$dh = -dz \tag{8.3-6}$$

Substitute equations (8.3–5) and (8.3–6) into equation (8.3–4) and write

$$dQ = -b\sqrt{(2g)}h^{\frac{1}{2}}\,dh \tag{8.3-7}$$

For a rectangular conduit, b is constant and equation (8.3–7) can be integrated to

$$Q = -\tfrac{2}{3}b\sqrt{(2g)}h^{\frac{3}{2}} + c \tag{8.3-8}$$

Substitute equation (8.3–5) into equation (8.3–8) and evaluate between the limits $z = 0$ and $z = z_o$ to give

$$Q = \frac{2}{3}b\sqrt{(2g)}\left[\left(z_o + \frac{v_o^2}{2g}\right)^{\frac{3}{2}} - \left(\frac{v_o^2}{2g}\right)^{\frac{3}{2}}\right] \tag{8.3-9}$$

If the approach velocity v_o is neglected equation (8.3–9) becomes

$$Q = \tfrac{2}{3}b\sqrt{(2g)}z_o^{\frac{3}{2}} \tag{8.3-10}$$

Equations (8.3–9) and (8.3–10) give the volumetric flow rate Q through a rectangular weir when there is no friction in the system.

In practice the volumetric flow rate Q is given by equation (8.3–11)

$$Q = \tfrac{2}{3}C_d b\sqrt{(2g)}z_o^{\frac{3}{2}} \qquad (8.3\text{–}11)$$

C_d is the dimensionless discharge coefficient which is a function of z_o and the ratio z_w/z_o where z_w is the height of the weir crest above the channel bed. A typical value is $C_d = 0.65$ for $z_w/z_o = 2$ and $z_o = 0.3$ m. Viscosity has a negligible influence on C_d.

In addition to rectangular weirs, V-notch or triangular weirs are commonly used with a cross-sectional flow area as shown in

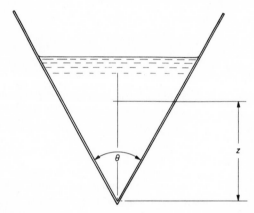

Figure (8.3–2)
Cross-sectional flow area in a V-notch weir.

Figure (8.3–2). In this case the width of the conduit b is variable and at any height z above the bottom of the weir is

$$b = 2z \tan \frac{\theta}{2} \qquad (8.3\text{–}12)$$

If the approach velocity v_o is neglected, equation (8.3–4) can be written

$$dQ = b\sqrt{(2g)}\sqrt{(z_o - z)}\,dz \qquad (8.3\text{–}13)$$

where dQ is the volumetric flow rate through an element of cross-sectional flow area of height dz when friction is neglected.

Substitute equation (8.3–12) into equation (8.3–13) and write

$$dQ = 2 \tan \frac{\theta}{2}\sqrt{(2g)}\sqrt{(z_o - z)z}\,dz \qquad (8.3\text{–}14)$$

Let

$$h = z_o - z \qquad (8.3\text{--}15)$$

so that

$$dh = -dz \qquad (8.3\text{--}6)$$

Substitute equations (8.3–6) and (8.3–15) into equation (8.3–14) and write

$$dQ = 2 \tan \frac{\theta}{2} \sqrt{(2g)} h^{\frac{1}{2}} (z_o - h) \, dz \qquad (8.3\text{--}16)$$

Integrate to

$$Q = 2 \tan \frac{\theta}{2} \sqrt{(2g)} (\tfrac{2}{3} z_o h^{\frac{3}{2}} - \tfrac{2}{5} h^{\frac{5}{2}}) + C \qquad (8.3\text{--}17)$$

Substitute equation (8.3–15) into equation (8.3–17) and evaluate between the limits $z = 0$ and $z = z_o$ to give

$$Q = \frac{8}{15} \tan \frac{\theta}{2} \sqrt{(2g)} z_o^{\frac{5}{2}} \qquad (8.3\text{--}18)$$

Equation (8.3–18) gives the volumetric flow rate Q through a V-notch weir when there is no friction in the system.

In practice the volumetric flow rate Q is given by Equation (8.3–19).

$$Q = \frac{8}{15} C_d \tan \frac{\theta}{2} \sqrt{(2g)} z_o^{\frac{5}{2}} \qquad (8.3\text{--}19)$$

where C_d is the dimensionless discharge coefficient which is mainly a function of z_o and θ. A typical value is $C_d = 0.62$ for $z_o = 0.15 \, \text{m}$ and $\theta = 20°$.

8.4 Mechanical and electromagnetic flowmeters

In the head flowmeters discussed so far, the primary element has no moving parts. The constriction area is fixed and the pressure drop varies as the flow rate changes. In the rotameter shown in Figure (8.4–1), the pressure drop is held constant and the constriction area varies as the flow rate changes. A float is free to move up and down in a tapered tube. The float remains steady when the upward force of the flowing fluid exactly balances the weight of the float in

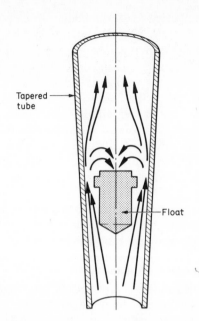

Figure (8.4–1)
Rotameter.

the fluid. The tapered tube is marked with a scale which is calibrated for a given fluid to give flow velocity at each scale reading. As the fluid flow rate is increased, the float moves to a higher position in the tube.

Other commonly used mechanical flowmeters are velocity meters. These usually consist of a non-magnetic casing, a rotor, and an electromagnetic pickup. The rotor is either a propeller or turbine freely suspended on ball bearings in the path of the flowing fluid with the axis of rotation in line with the flow. The rotor turns in the fluid flow stream at a rate proportional to the flow rate. As the rotor turns it cuts through the lines of force of an electric field produced by an adjacent induction coil. The electrical pulse output from the induction coil pickup is amplified and fed to readout instruments or recorders to give either total flow or flow rate.[5]

The head flowmeters and mechanical flowmeters considered so far all involve some kind of restriction in the flow line, which in turn produces additional frictional head losses. In contrast, the electromagnetic flowmeter consists of a straight length of non-magnetic pipe containing no restrictions through which the fluid flows. The pipe is normally lined with electrically insulating material.

Two small electrodes located diametrically opposite each other are sealed in flush with the interior surface. Coil windings on the outside of the non-magnetic pipe provide a magnetic field. The flowing fluid acts as a moving conductor which cuts the magnetic lines of force. A voltage is induced in the fluid which is directly proportional to the fluid velocity. This voltage is detected by the electrodes and is then amplified and transmitted to either readout instruments or recorders. The magnetic field is usually alternating so that the induced voltage is also alternating. Electromagnetic flowmeters can only be used to meter fluids which have some electrical conductivity. They cannot be used to meter hydrocarbons.

Although electromagnetic flowmeters are expensive they are especially suitable for metering liquids containing suspended solids. Furthermore, unlike head flowmeters, they are unaffected by variations in fluid viscosity, density or temperature. Since they are also unaffected by turbulence or variations in velocity profile, they can be installed close to valves, bends, fittings, etc.

The flowmeters discussed above are used either to measure linear velocity or volumetric flow rate. They can only be used to measure flow rate in mass per unit time if the fluid density is also measured and the volumetric flow rate and density signals are coordinated.

Direct reading mass flowmeters have now been developed. These operate on the angular momentum principle. The primary element consists of a cylindrical impeller and turbine both of which contain tubes through which the fluid flows. Angular momentum is given to the fluid by driving the impeller at a constant speed via a magnetic coupling using a synchronous motor. Fluid from the impeller discharges into the cylindrical turbine where all the angular momentum is removed. The torque on the turbine is proportional to the flow rate in mass per unit time. This torque is transferred to a gyro-integrating mechanism via a magnetic coupling. A cyclometer registers the rotation of the gyroscope and totalizes the fluid flow rate in mass per unit time. Direct reading mass flowmeters are expensive.

8.5   Scale errors in flow measurement

Rotameters, velocity flowmeters, and electromagnetic flowmeters have the advantage that they can be used with a linear scale in which the volumetric flow rate Q is directly proportional to the scale reading s.

$$Q = k_1 s \qquad (8.5\text{--}1)$$

where k_1 is a constant.

Differentiate equation (8.5–1) to give

$$\frac{\mathrm{d}Q}{\mathrm{d}s} = k_1 \qquad (8.5\text{--}2)$$

All head flowmeters are used with a square root scale in which the volumetric flow rate Q is proportional to the square root of the scale reading s.

$$Q = k_2 s^{\frac{1}{2}} \qquad (8.5\text{--}3)$$

where k_2 is a constant.

Differentiate equation (8.5–3) to give

$$\frac{\mathrm{d}Q}{\mathrm{d}s} = \frac{k_2}{2s^{\frac{1}{2}}} = \frac{k_2^2}{2Q} \qquad (8.5\text{--}4)$$

Buzzard[3] defined the per cent flow rate error Δe as

$$\Delta e = 100\frac{\Delta Q}{Q} \qquad (8.5\text{--}5)$$

where ΔQ is the absolute error in the volumetric flow rate.
Let Δs be the indicator or recorder error.
Write

$$\frac{\Delta Q}{\Delta s} \simeq \frac{\mathrm{d}Q}{\mathrm{d}s} \qquad (8.5\text{--}6)$$

Combine equations (8.5–5) and (8.5–6) and write

$$\Delta e = \left(\frac{100\,\Delta s}{Q}\right)\left(\frac{\mathrm{d}Q}{\mathrm{d}s}\right) \qquad (8.5\text{--}7)$$

Substitute equation (8.5–2) for a linear scale into equation (8.5–7) to give

$$\Delta e = \frac{100\,\Delta s\,k_1}{Q} \qquad (8.5\text{--}8)$$

If the maximum volumetric flow rate $Q_{max} = 100$ and the maximum scale reading $s_{max} = 100$, equation (8.5–8) can be written as

$$\Delta e = \frac{100 \, \Delta s}{Q} \tag{8.5–9}$$

Substitute equation (8.5–4) for a square root scale into equation (8.5–7) to give

$$\Delta e = \frac{50 \, \Delta s \, k_2^2}{Q^2} \tag{8.5–10}$$

If the maximum volumetric flow rate $Q_{max} = 100$ and the maximum scale reading $s_{max} = 100$, equation (8.5–10) can be written as

$$\Delta e = \frac{5000 \, \Delta s}{Q^2} \tag{8.5–11}$$

The square root scale is more accurate than the linear scale at flow rates near the maximum of the scale. The linear scale is more accurate than the square root scale at flow rates much less than the maximum of the scale. Thus head flowmeters are unsuitable for measuring flow rates which vary widely.

Example (8.5–1)

A flowmeter is inherently accurate at all points to 0.5 per cent of the full range. Calculate the per cent flow rate error using first a linear and then a square root scale for flow rates of 10, 25, 50 and 100 per cent of maximum flow.

Calculations:
  indicator error

$$\Delta s = 0.5$$

if $s_{max} = 100$

for a linear scale, per cent flow rate error

$$\Delta e = \frac{100 \, \Delta s}{Q} \tag{8.5–9}$$

for $Q_{max} = 100$ and $s_{max} = 100$,

therefore

$$\Delta e = \frac{(100)(0.5)}{10} = 5 \text{ per cent for a flow rate 10 per cent of } Q_{max}$$

$$\Delta e = \frac{(100)(0.5)}{25} = 2 \text{ per cent for a flow rate 25 per cent of } Q_{max}$$

$$\Delta e = \frac{(100)(0.5)}{50} = 1 \text{ per cent for a flow rate 50 per cent of } Q_{max}$$

$$\Delta e = \frac{(100)(0.5)}{100} = 0.5 \text{ per cent for } Q_{max}$$

for a square root scale, per cent flow rate error

$$\Delta e = \frac{5000 \, \Delta s}{Q^2} \qquad (8.5\text{--}11)$$

for $Q_{max} = 100$ and $s_{max} = 100$,

therefore

$$\Delta e = \frac{(5000)(0.5)}{(10)^2} = 25 \text{ per cent for a flow rate 10 per cent of } Q_{max}$$

$$\Delta e = \frac{(5000)(0.5)}{(25)^2} = 4 \text{ per cent for a flow rate 25 per cent of } Q_{max}$$

$$\Delta e = \frac{(5000)(0.5)}{(50)^2} = 1 \text{ per cent for a flow rate 50 per cent of } Q_{max}$$

$$\Delta e = \frac{(5000)(0.5)}{(100)^2} = 0.25 \text{ per cent for } Q_{max}$$

REFERENCES

(1) Barna, P. S., *Fluid Mechanics for Engineers*, p. 109, London, Butterworth and Company (Publishers) Ltd., 1969.
(2) Ibid., p. 112.
(3) Buzzard, W., *Chem. Eng.*, **66**, No. 5 (1959).
(4) Foust, A. S., Wenzel, L. A., Clump, C. W., Maus, L., and Anderson, L. B., *Principles of Unit Operations*, p. 23, New York, John Wiley and Sons, Inc., 1964.
(5) Holland, F. A., and Chapman, F. S., *Pumping of Liquids*, p. 280, New York, Reinhold Publishing Corporation, 1966.
(6) Ziemke, M. C., and McCallie, B. G., Chem. Eng., **71**, No. 19 (1964).

9
Fluid motion in the presence of solid particles

9.1 Relative motion between a fluid and a single particle

Consider the relative motion between a particle and an infinitely large volume of fluid. Since only the relative motion is considered the following cases are covered:
(1) a stationary particle in a moving fluid;
(2) a moving particle in a stationary fluid;
(3) a particle and a fluid moving in opposite directions;
(4) a particle and a fluid both moving in the same direction but at different velocities.

Consider the flow past a spherical particle shown in Figure (9.1–1). The fluid is brought to rest at points 1 and 2. From the Bernoulli equation (equation (1.6–4)), it is seen that the pressure is a maximum at points 1 and 2. Similarly since the velocity is a maximum at

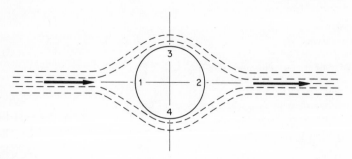

Figure (9.1–1)
Flow past a spherical particle.

points 3 and 4, these are the points of minimum pressure. As the relative velocity between the particle and the fluid increases, the pressure at point 2 increases. This causes the boundary layer to thicken and separate from the surface. Eddies are formed in the wake of the particle with a consequent dissipation of energy. The resultant force on the particle is known as form drag.

At very low velocities, no separation of the boundary layer from the particle surface occurs. In this case, the only force on the particle is that due to skin friction. As a result of skin friction, the fluid in immediate contact with the particle is at rest. At higher velocities, the force on the particle results from a combination of skin friction and form drag.

Consider a spherical particle of diameter d_p and density ρ_p falling steadily with a linear velocity u_p under the influence of gravity in a fluid of density ρ. The gravitational force F_1 on the particle is given by the equation

$$F_1 = \frac{\pi d_p^3}{6}(\rho_p - \rho)g \qquad (9.1\text{--}1)$$

where $\pi d_p^3/6$ is the volume of the spherical particle.

The retarding force F_2 on the particle from the fluid is given by the equation

$$F_2 = C_d S_p \frac{\rho u_p^2}{2} \qquad (9.1\text{--}2)$$

where C_d is a dimensionless drag coefficient and S_p is the projected area of the particle in a plane perpendicular to the direction of the fluid stream. Equation (9.1–2) may be obtained by dimensional analysis.

For steady flow the forces F_1 and F_2 are equal and opposite and equations (9.1–1) and (9.1–2) can be combined and written as

$$\frac{\pi d_p^3}{6}(\rho_p - \rho)g = C_d S_p \frac{\rho u_p^2}{2} \qquad (9.1\text{--}3)$$

For a spherical particle $S_p = \pi d_p^2/4$ and for this case equation (9.1–3) can be rewritten in the form

$$u_p = \sqrt{\frac{4d_p(\rho_p - \rho)g}{3C_d\rho}} \qquad (9.1\text{--}4)$$

where u_p is known as the terminal settling or falling velocity.

The dimensionless drag coefficient C_d is a function of a Reynolds number defined as

$$(N_{RE})_p = \frac{\rho u_p d_p}{\mu} \qquad (9.1\text{-}5)$$

where ρ and μ are the density and dynamic viscosity of the fluid respectively.

For the streamline Reynolds number range $(N_{RE})_p < 0.2$

$$C_d = \frac{24}{(N_{RE})_p} \qquad (9.1\text{-}6)$$

Combine equations (9.1–5) and (9.1–6) and substitute into equation (9.1–4) to give

$$u_p = \frac{d_p^2(\rho_p - \rho)g}{18\mu} \qquad (9.1\text{-}7)$$

Equation (9.1–7) was derived theoretically by Stokes[17] and is known as the Stokes equation for steady streamline flow past a sphere.

For the Reynolds number range $0.2 < (N_{RE})_p < 500$, Schiller and Naumann[15] have shown that

$$C_d = \frac{24}{(N_{RE})_p}[1 + 0.15(N_{RE})_p^{0.687}] \qquad (9.1\text{-}8)$$

Equation (9.1–8) is an empirical equation which is only approximately true.

For the Reynolds number range $500 < (N_{RE})_p < 200\,000$

$$C_d = 0.44 \qquad (9.1\text{-}9)$$

For the most part, solid particles in fluid streams have Reynolds numbers which are much lower than 500.

Pettyjohn and Christiansen[13] gave equations for the terminal settling velocities of particles which deviate from a spherical shape.

Lapple and Shepherd[7,10] presented plots of the dimensional drag coefficient C_d against the particle Reynolds number $(N_{RE})_p$ for spheres, discs and cylinders.

Equation (9.1–4) gives the terminal settling velocity for a spherical particle. For a non-spherical particle, equation (9.1–4) can be written in the modified form

$$u_p = \sqrt{\frac{4d_p\psi(\rho_p - \rho)g}{3C_d\rho}} \qquad (9.1\text{-}10)$$

For a spherical particle the dimensionless correction factor $\psi = 1$ and equations (9.1–4) and (9.1–10) become identical.

9.2   Relative motion between a fluid and a concentration of particles

So far the relative motion between a fluid and a single particle has been considered. This process is called free settling. When a fluid contains a concentration of particles in a vessel, the settling of an individual particle may be hindered by the other particles and by the walls. When this is the case, the process is called hindered settling. Interference is negligible if the particles are at least 10 to 20 diameters away from each other and the vessel wall.[8] In this case the particles can be considered to be free settling.

Hindered settling results from collisions between particles and also between particles and the wall. In addition high particle concentrations reduce the flow area and increase the linear velocity of the fluid with a consequent decrease in settling rate. Furthermore particle concentrations increase the apparent density and dynamic viscosity of the fluid.

Richardson and Zaki[14] showed that in the Reynolds number range $(N_{RE})_p < 0.2$, the velocity u_c of a suspension of coarse spherical particles in water relative to a fixed horizontal plane is given by the equation

$$\frac{u_c}{u_p} = \varepsilon^{4.6} \qquad (9.2-1)$$

where u_p is the terminal settling velocity for a single particle and ε is the voidage fraction of the suspension which is unity for a single particle in an infinite amount of fluid.

Settling can be used to classify or separate particles since different sized particles settle at different velocities. Similarly elutriation can also be used to classify particles where small particles are carried upwards with the fluid and large particles sink. Particles of the same size but of different densities can also be separated by settling or elutriation.

Consider two spherical particles 1 and 2 of the same diameter but of different densities settling freely in a fluid of density ρ in the streamline Reynolds number range $(N_{RE})_p < 0.2$. The ratio of the terminal settling velocities u_{p1}/u_{p2} is given by equation (9.1–7) rewritten in the form

$$\frac{u_{p1}}{u_{p2}} = \frac{\rho_{p1} - \rho}{\rho_{p2} - \rho} \qquad (9.2-2)$$

The greater the ratio u_{p1}/u_{p2} the greater the ease of separation. Thus the fluid density ρ can be chosen to give a high ratio of terminal settling velocities.

Consider two spherical particles 1 and 2 of the same density ρ_p but of different diameters settling freely in a fluid of density ρ in the streamline Reynolds number range $(N_{RE})_p < 0.2$. The ratio of the terminal settling velocities u_{p1}/u_{p2} is given by equation (9.1–7) rewritten in the form

$$\frac{u_{p1}}{u_{p2}} = \left(\frac{d_{p1}}{d_{p2}}\right)^2 \qquad (9.2\text{–}3)$$

where d_{p1}/d_{p2} is the ratio of the particle diameters.

Consider two spherical particles 1 and 2 of different densities and diameters settling freely in a fluid of density ρ with the same terminal settling velocity in the streamline Reynolds number range $(N_{RE})_p < 0.2$. The densities and diameters are related by equation (9.1–7) rewritten in the form

$$\frac{d_{p1}}{d_{p2}} = \sqrt{\left|\frac{\rho_{p2} - \rho}{\rho_{p1} - \rho}\right|} \qquad (9.2\text{–}4)$$

The classification or separation of particles can be carried out more rapidly in centrifugal separators than in gravity settlers. In gravity settlers, the particles travel vertically downwards, whereas in centrifugal separators the particles travel radially outwards. A particle of mass m rotating at a radius r with an angular velocity ω is subject to a radially directed centrifugal force $mr\omega^2$ which can be made very much greater than the vertically directed gravity force mg.

The terminal settling velocity u_p for a single spherical particle in a centrifugal separator can be calculated from equation (9.1–4) with the centrifugal acceleration $r\omega^2$ replacing the gravitational acceleration g to give

$$u_p = \sqrt{\frac{4d_p(\rho_p - \rho)r\omega^2}{3C_d\rho}} \qquad (9.2\text{–}5)$$

A very small particle may still be in laminar flow in a centrifugal separator. In this case $r\omega^2$ may be written in place of g in equation (9.1–7) to give

$$u_p = \frac{d_p^2(\rho_p - \rho)r\omega^2}{18\mu} \qquad (9.2\text{–}6)$$

In addition to hydrodynamic interactions between solid particles and a fluid, physico-chemical forces may act between pairs of particles. These forces tend to form a structure which prevents the particles from settling out.[2] If the forces are sufficiently strong a homogeneous slurry results which usually has non-Newtonian rheological characteristics. If the structure is weak, the slurry is pseudoplastic. Slurries with a high proportion of solids tend to be dilatant.

Einstein[6] studied homogeneous slurries of spherical particles in a liquid of the same density. He showed that the distortion of the streamlines around the particles caused the dynamic viscosity of the slurry to increase according to the equation

$$\mu = \mu_L(1 + 2.5x_v) \qquad (9.2\text{–}7)$$

where μ and μ_L are the dynamic viscosities of the slurry and liquid respectively and x_v is the volume concentration of the solids. Equation (9.2–7) holds for low concentrations up to $x_v = 0.2$.

9.3 Fluid flow through packed beds

In a packed bed of unit volume, the volumes occupied by the voids and the solid particles are ε and $(1 - \varepsilon)$ respectively where ε is the voidage fraction or porosity of the bed. Let S_o be the surface area per unit volume of the solid material in the bed. Thus the total surface area in a packed bed of unit volume is $(1 - \varepsilon)S_o$.

For a spherical particle of diameter d_p the value of $S_o = 6/d_p$. For a non-spherical particle with an average particle diameter d_p, the value $S_o = 6/(d_p\psi)$ where $\psi = 1$ for a spherical particle. Values of ψ for other shapes are readily available.[10]

An equivalent diameter d_e for flow through the bed can be defined as four times the cross-sectional flow area divided by the appropriate flow perimeter. Thus if the packed bed considered has unit height, the equivalent diameter is

$$d_e = \frac{4\varepsilon}{(1 - \varepsilon)S_o} \qquad (9.3\text{–}1)$$

If a fluid is flowing steadily with a mean linear approach velocity u, the apparent mean linear velocity in the packed bed $u_b = u/\varepsilon$.

A Reynolds number for flow through a packed bed can be defined as

$$(N_{RE})_b = \frac{\rho u_b d_e}{\mu} \qquad (9.3\text{--}2)$$

which when combined with equation (9.3–1) can be written as

$$(N_{RE})_b = \frac{4\rho u}{\mu(1 - \varepsilon)S_o} \qquad (9.3\text{--}3)$$

An alternative Reynolds number has been used to correlate data and is defined as

$$(N_{RE})'_b = \frac{\rho u}{\mu(1 - \varepsilon)S_o} \qquad (9.3\text{--}4)$$

For a packed bed consisting of spherical particles, equation (9.3–4) can be written in the form

$$(N_{RE})'_b = \frac{\rho u d_p}{6\mu(1 - \varepsilon)} \qquad (9.3\text{--}5)$$

The corresponding equation for non-spherical particles is

$$(N_{RE})'_b = \frac{\rho u d_p \psi}{6\mu(1 - \varepsilon)} \qquad (9.3\text{--}6)$$

Consider fluid flowing steadily through a packed bed of height L and unit cross-sectional area. A pressure ΔP occurs in the bed because of frictional viscous and drag forces. Let the resistance per unit area of surface be R_b. A force balance across unit cross-sectional area gives

$$\Delta P\, \varepsilon = R_b L(1 - \varepsilon)S_o \qquad (9.3\text{--}7)$$

which can be written either as

$$(j_f)_b = \frac{R_b}{\rho u_b^2} = \left(\frac{\Delta P}{L}\right)\left[\frac{\varepsilon}{(1 - \varepsilon)S_o \rho u_b^2}\right] \qquad (9.3\text{--}8)$$

or since $u_b = u/\varepsilon$ as

$$(j_f)_b = \frac{R_b}{\rho u_b^2} = \left(\frac{\Delta P}{L}\right)\left[\frac{\varepsilon^3}{(1 - \varepsilon)S_o \rho u^2}\right] \qquad (9.3\text{--}9)$$

where $(j_f)_b$ is a dimensionless friction factor for flow through a packed bed.

For laminar flow where $(N_{RE})'_b \leqslant 2$

$$(j_f)_b = \frac{5}{(N_{RE})'_b} \qquad (9.3-10)$$

The transition to turbulent flow is gradual. Turbulence commences initially in the largest channels and eventually extends to the smaller channels.

For the complete range of Reynolds number, Carmen[1] gave the equation

$$(j_f)_b = \frac{5}{(N_{RE})'_b} + \frac{0.4}{[(N_{RE})'_b]^{0.1}} \qquad (9.3-11)$$

Log log plots of $(j_f)'_b$ against $(N_{RE})_b$ are readily available[11] for randomly packed beds.

The Hagen–Poiseuille equation for steady laminar flow of Newtonian fluids in pipes and tubes can be written as

$$u = \left(\frac{\Delta P}{L}\right)\frac{d_i^2}{32\mu} \qquad (2.4-6)$$

For a packed bed, substitute the equivalent diameter d_e from equation (9.3–1) into equation (2.4–6) to give

$$u_b = \left(\frac{\Delta P}{L}\right)\left(\frac{1}{32\mu}\right)\left[\frac{16\varepsilon^2}{(1-\varepsilon)^2 S_o^2}\right] \qquad (9.3-12)$$

or since $u_b = u/\varepsilon$ to give

$$u = \left(\frac{\Delta P}{L}\right)\left(\frac{1}{2\mu}\right)\left[\frac{\varepsilon^3}{(1-\varepsilon)^2 S_o^2}\right] \qquad (9.3-13)$$

Equation (9.3–13) does not hold for flow through packed beds and should be replaced by the equation

$$u = \left(\frac{\Delta P}{L}\right)\left(\frac{1}{K_c\mu}\right)\left[\frac{\varepsilon^3}{(1-\varepsilon)^2 S_o^2}\right] \qquad (9.3-14)$$

which can also be written in the form

$$\Delta P = (K_c\mu L)\left[\frac{(1-\varepsilon)^2 S_o^2}{\varepsilon^3}\right]u \qquad (9.3-15)$$

Equation (9.3–14) is the Carmen Kozeny equation[1]. The parameter K_c has a value which depends on the particle shape, the porosity and particle size range. The value lies in the range 3.5 to 5.5 but the value most commonly used is 5.

For spherical particles $S_o = 6/d_p$ and equation (9.3–15) can be written as

$$\Delta P = (180\mu L)\left[\frac{(1 - \varepsilon)^2}{\varepsilon^3 d_p^2}\right]u \qquad (9.3\text{–}16)$$

where $K_c = 5.0$.

Example (9.3–1)

A gas of density 1.25 kg/m^3 and dynamic viscosity 1.5×10^{-5} kg/(s m) flows steadily through a bed of spherical particles 0.005 m in diameter. The bed has a height of 3.00 m and a voidage of $\frac{1}{3}$. The linear approach velocity is 0.03 m/s. Calculate the Reynolds number and the pressure drop in N/m^2 over the bed.

Calculations:

$$\text{Reynolds number } (N_{RE})'_b = \frac{\rho u d_p}{6\mu(1 - \varepsilon)} \qquad (9.3\text{–}5)$$

$$\rho = 1.25 \text{ kg/m}^3$$

$$u = 0.03 \text{ m/s}$$

$$d_p = 0.005 \text{ m}$$

$$\mu = 1.5 \times 10^{-5} \text{ kg/(s m)}$$

$$\varepsilon = \tfrac{1}{3}$$

$$(1 - \varepsilon) = \tfrac{2}{3}$$

$$(N_{RE})'_b = \frac{(1.25 \text{ kg/m}^3)(0.03 \text{ m/s})(0.005 \text{ m})(3)}{(6)[1.25 \times 10^{-5} \text{ kg/(s m)}](2)}$$

$$= 3.75$$

$$\text{pressure drop } \Delta P = (180\mu L)\left[\frac{(1 - \varepsilon)^2}{\varepsilon^3 d_p^2}\right]u \qquad (9.3\text{–}16)$$

$$(1 - \varepsilon)^2 = \tfrac{4}{9}$$

$$\frac{(1 - \varepsilon)^2}{\varepsilon^3} = 12$$

$$d_p^2 = 2.5 \times 10^{-5} \text{ m}^2$$

$$u = 0.03 \text{ m/s}$$
$$\mu = 1.5 \times 10^{-5} \text{ kg/(s m)}$$
$$L = 3.0 \text{ m}$$

$$\Delta P = (180)[1.5 \times 10^{-5} \text{ kg/(s m)}]\frac{(3.0 \text{ m})(12)(0.03 \text{ m/s})}{(2.5 \times 10^{-5} \text{ m}^2)}$$

$$= 116.6 \text{ N/m}^2$$

9.4 Fluidisation

If a fluid in laminar flow is passed upwards through a static packed bed of solid particles the pressure gradient is given by equation (9.3–14)

$$u = \left(\frac{\Delta P}{L}\right)\left(\frac{1}{K_c \mu}\right)\left[\frac{\varepsilon^3}{(1-\varepsilon)^2 S_o^2}\right] \qquad (9.3–14)$$

As the fluid velocity is increased a point is reached when the viscous frictional and drag forces on the particles become equal to the weight of the particles in the fluid stream. This is the start of fluidisation and a force balance gives

$$\Delta P = (1-\varepsilon)(\rho_p - \rho)Lg \qquad (9.4–1)$$

where ρ_p and ρ are the densities of the solid and fluid respectively.

Combine equations (9.3–14) and (9.4–1) to give

$$u = \left[\frac{(\rho_p - \rho)g}{K_c \mu}\right]\left[\frac{\varepsilon^3}{(1-\varepsilon)S_o^2}\right] \qquad (9.4–2)$$

where K_c is usually about 5.

For spherical particles $S_o = 6/d_p$ and equation (9.4–2) can be written

$$u = \left[\frac{(\rho_p - \rho)g}{\mu}\right]\left[\frac{d_p^2 \varepsilon^3}{180(1-\varepsilon)}\right] \qquad (9.4–3)$$

where in this case ε is the porosity at the point of fluidisation.

If the velocity is further increased, the bed expands. This type of fluidisation is known as particulate fluidisation and is obtained at all liquid velocities and at low gas velocities. At higher gas velocities, regions of gas containing some solid particles bubble through the fluidised bed in a manner similar to gas bubbles in a boiling liquid.

This type of fluidisation is called aggregative fluidisation or boiling and the bed as a boiling bed.

As the fluid velocity is increased, the bed expands and solid particles become entrained. Initially the smaller particles only are carried away. If the fluid velocity is sufficiently increased, all the particles will become entrained.

For a spherical particle of diameter d_p to be entrained, the fluid velocity must exceed the terminal falling velocity given by equation (9.1–4)

$$u_p = \sqrt{\frac{4d_p(\rho_p - \rho)g}{3C_d\rho}} \qquad (9.1\text{–}4)$$

where C_d is the dimensionless drag coefficient for the particle.

The mean linear velocity of fluid in a fluidised bed of spherical particles must have a value which is greater than the fluidisation velocity given by equation (9.4–3) and less than the entrainment velocity given by equation (9.1–4).

9.5   Slurry transport

A slurry is a liquid containing solid particles in suspension. Slurries can be divided into two classes: settling and non-settling.

Non-settling slurries usually consist of a high concentration of finely divided solid particles suspended in a liquid. The solid particles may also settle so slowly that the slurry may be regarded for all practical purposes as non-settling. Like true liquids, non-settling slurries may exhibit either Newtonian or non-Newtonian flow behaviour. Milk is a non-settling slurry which behaves as a thixotropic liquid. Non-settling homogeneous slurries can be pumped through a pipeline either in laminar or turbulent flow.

Compared with non-settling slurries, settling slurries contain larger solid particles at lower concentrations. Settling slurries are essentially two-phase heterogeneous mixtures. The liquid and the solid particles exhibit their own characteristics. Thus in contrast to non-settling slurries, the solid particles in settling slurries do not alter the viscosity of the conveying liquid.

Settling slurries cannot be pumped in laminar flow. Turbulence must exist to prevent the solid particles from settling. Settling slurries should be pumped through pipelines at velocities which just prevent the solid particles from settling. This results in the minimum pressure drop across the pipeline.

Saltation[3] is also used to transport settling slurries through pipelines. In this case the solid particles bounce and roll along the bottom of a horizontal pipe.

Below a certain minimum velocity, the turbulence is insufficient to keep all the particles suspended in a settling slurry flowing through a horizontal pipe. At this minimum velocity, there is a concentration gradient from the top to the bottom of the horizontal pipe. At a higher velocity called the standard velocity this concentration disappears and the flow becomes homogeneous. Spells[16] called the region between the minimum and standard velocities, the heterogeneous flow region.

Empirical equations are available[5] which predict values for the minimum and standard velocities for various slurries. Spells[16] analysed the experimental data of a number of investigators for aqueous slurries of sands, boiler ash and lime flowing in horizontal pipes. He obtained the following empirical equations which give the mean minimum linear liquid velocity u_1 and the mean standard linear liquid velocity u_2 respectively for slurries in horizontal pipes:

$$u_1 = \left[0.0251 g d_p \left(\frac{\rho_m d_i}{\mu} \right)^{0.775} \left(\frac{\rho_p - \rho}{\rho} \right) \right]^{1/1.225} \qquad (9.5\text{--}1)$$

and

$$u_2 = \left[0.0741 g d_p \left(\frac{\rho_m d_i}{\mu} \right)^{0.775} \left(\frac{\rho_p - \rho}{\rho} \right) \right]^{1/1.225} \qquad (9.5\text{--}2)$$

Equations (9.5–1) and (9.5–2) are based on experimental data for solid particles with diameters in the range 6×10^{-5} to 6×10^{-4} m and pipe diameters of 2.5×10^{-2} to 3×10^{-1} m. In equations (9.5–1) and (9.5–2) μ is the dynamic viscosity of the transporting liquid and ρ, ρ_p, and ρ_m are the densities of the liquid, solid particles and slurry mixture respectively. The latter is given by the equation

$$\rho_m = x_v(\rho_s - \rho) + \rho \qquad (9.5\text{--}3)$$

where x_v is the volume fraction of the solids in the slurry.

Empirical equations[4] are also available to calculate the pressure drop for slurries flowing through pipelines. Durand and Condolios[5] found the following equation to fit the experimental data for sand water mixtures flowing above the minimum velocity in horizontal pipes:

$$\frac{\Delta P - \Delta P_W}{x_v \, \Delta P_W} = \frac{180}{\{[u^2/(gd_i)]C_d^{\frac{1}{2}}\}^{\frac{3}{2}}} \qquad (9.5\text{--}4)$$

In equation (9.5–4), ΔP and ΔP_W are the pressure drops for the slurry and for the clear water respectively.

Condolios and Chapus[4] found that the presence of fines in a coarse slurry decreases the frictional pressure drop in a horizontal pipe to a much greater extent than might be expected from their relative properties in the solids.

Newitt, Richardson, and Gliddon[9] carried out experiments on aqueous slurries of pebbles, zircon, manganese dioxide, perspex and various kinds of sand in vertical pipes of 2.5×10^{-3} m and 5.0×10^{-3} m diameter. Their pressure drop data were satisfactorily correlated with the following equation:

$$\frac{\Delta P - \Delta P_W}{x_v \, \Delta P_W} = 0.0037 \left(\frac{g d_i}{u^2}\right)^{\frac{1}{2}} \left(\frac{d_i}{d_p}\right) \left(\frac{\rho_p}{\rho}\right)^2 \qquad (9.5\text{–}5)$$

In equation (9.5–5), the mean linear velocity u is the volumetric flow rate of the slurry divided by the cross-sectional area of the pipe.

Solid particles hydraulically conveyed in a vertical pipe have a mean linear velocity which is less than the mean linear velocity of the liquid. This is because of the tendency of the particles to settle. The volume fraction x_v in equation (9.5–5) is the delivered concentration. This is less than the volume fraction in the vertical pipe.

Solid particles hydraulically conveyed in a vertical pipe are subjected to forces which cause them to rotate and move inwards towards the axis of the pipe. This is known as the Magnus effect and is most pronounced with large velocity gradients.

9.6   Filtration

When a slurry flows through a filter, the solid particles become entrapped by the filter medium which is permeable only to the liquid. Either of two mechanisms are used: cake filtration or depth filtration.

In cake filtration, the filter medium acts as a strainer and collects the solid particles on top of the initial layer. A filter cake is formed and the flow obeys the Carmen Kozeny equation for packed beds.

Depth filtration is also called granular filtration. In this case the filter medium is a bed of particulate material through which the slurry flows. Solid particles in the slurry are carried right into and are deposited within the bed. The bed is deep compared to its grain size. The latter is also much larger than the grain size in the slurry.

There is virtually no deposition on the surface of the bed. Granular filters are suitable for producing high quality filtrate from large quantities of liquid containing up to 50 parts per million solids. The performance depends not only on the minimum particle size to be removed but also on the affinity of the suspended particles for the granular material. The most commonly used granular material is silica sand.

REFERENCES

(1) Carmen, P. C., Trans. Inst. Chem. Eng., **15**, 150 (1937).
(2) Cheng, D. C. H., Filtration and Separation, 7, 434 (1970).
(3) Condolios, E., and Chapus, E. E., Chem. Eng., **70**, No. 13 (1963).
(4) Condolios, E., and Chapus, E. E., Chem. Eng., **70**, No. 14 (1963).
(5) Durand, R., and Condolios, E., Centenary Congr. Mineral Industry, France, 1955.
(6) Einstein, A., Ann. Phys., **19**, 289 (1906).
(7) Lapple, C. E., and Shepherd, C. B., Ind. Eng. Chem., **32**, 605 (1940).
(8) Larian, M. G., *Fundamentals of Chemical Engineering Operations*, p. 542, Englewood Cliffs, N.J., Prentice-Hall Inc., 1958.
(9) Newitt, D. M., Richardson, J. F., and Gliddon, B. J., Trans. Inst. Chem. Eng., **39**, 93 (1961).
(10) Perry, J. H., *Chemical Engineers' Handbook*, p. 5–50, New York, McGraw-Hill Book Co. Inc., 1963.
(11) Ibid., p. 5–51.
(12) Ibid., p. 5–60.
(13) Pettyjohn, E. S., and Christiansen, E. B., Chem. Eng. Prog., **44**, 157 (1948).
(14) Richardson, J. F., and Zaki, W. W., Trans. Inst. Chem. Eng., **32**, 35 (1954).
(15) Schiller, L., and Naumann, A., Z. Ver. deut. Ing., **77**, 318 (1933).
(16) Spells, K. E., Trans. Inst. Chem. Eng., **33**, 79 (1955).
(17) Stokes, G. G., Trans. Cambridge Phil. Soc., **9**, 51 (1851).

10
Introduction to unsteady state fluid flow

10.1   Time to empty liquid from a tank

The Bernoulli equation applied to points A and B in the system shown in Figure (10.1–1) can be written as

$$z + \frac{P_A}{\rho g} + 0 = 0 + \frac{P_B}{\rho g} + \frac{u^2}{2g\alpha} + h_f \qquad (10.1\text{–}1)$$

or as

$$z + \frac{(P_A - P_B)}{\rho g} = \frac{u^2}{2g\alpha} + h_f \qquad (10.1\text{–}2)$$

In equation (10.1–2), h_f is the head loss due to friction in the outlet pipe and is given by the equation

$$h_f = 8j_f \left(\frac{\Sigma L_e}{d_i} \right) \frac{u^2}{2g} \qquad (10.1\text{–}3)$$

In equation (10.1–3) ΣL_e is the equivalent length of the outlet pipe and d_i is its inside diameter; u is the mean linear velocity in the outlet pipe. It is assumed that the mean linear velocity in the tank is so small that it can be neglected.

Combine equations (10.1–2) and (10.1–3) to give

$$z + \frac{(P_A - P_B)}{\rho g} = \frac{u^2}{2g} \left[\frac{1}{\alpha} + 8j_f \left(\frac{\Sigma L_e}{d_i} \right) \right] \qquad (10.1\text{–}4)$$

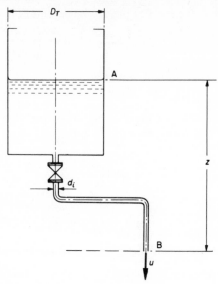

Figure (10.1–1)
Liquid flowing from a tank.

which can also be written as

$$u = \left\{ \frac{2g[z + (P_A - P_B)/(\rho g)]}{1/\alpha + 8j_f(\Sigma L_e/d_i)} \right\}^{\frac{1}{2}} \qquad (10.1\text{--}5)$$

Let the volume of liquid in the tank decrease by dV and the height of liquid in the tank decrease by dz in a time dt. Therefore

$$dV = \frac{\pi}{4} D_T^2 \, dz = \frac{\pi}{4} d_i^2 u \, dt \qquad (10.1\text{--}6)$$

which can also be written as

$$dt = \frac{D_T^2}{d_i^2} \frac{dz}{u} \qquad (10.1\text{--}7)$$

Substitute equation (10.1–5) into equation (10.1–7) to give

$$dt = \frac{D_T^2}{d_i^2} \left\{ \frac{1/\alpha + 8j_f(\Sigma L_e/d_i)}{2g[z + (P_A - P_B)/(\rho g)]} \right\}^{\frac{1}{2}} dz \qquad (10.1\text{--}8)$$

Equation (10.1–8) can be integrated to obtain the time for the liquid in the tank to fall from one level to another under the influence of gravity. Thus the time $\Delta t(1 - 2)$ for the liquid to fall from height z_1

to height z_2 above the level of the discharge end of the exit pipe is given by the equation

$$\Delta t(1 - 2) = \{[z_1 + (P_A - P_B)/(\rho g)]^{\frac{1}{2}} - [z_2 + (P_A - P_B)/(\rho g)]^{\frac{1}{2}}\}$$
$$\times \left(\frac{2D_T^2}{d_i^2}\right)\left[\frac{1/\alpha + 8j_f(\Sigma L_e/d_i)}{2g}\right]^{\frac{1}{2}} \quad (10.1-9)$$

If the pressures at points A and B are the same, equation (10.1–9) can be written in the simplified form as

$$\Delta t (1 - 2) = (z_1^{\frac{1}{2}} - z_2^{\frac{1}{2}})\left(\frac{2D_T^2}{d_i^2}\right)\left[\frac{1/\alpha + 8j_f(\Sigma L_e/d_i)}{2g}\right]^{\frac{1}{2}} \quad (10.1-10)$$

Example (10.1–1)

Calculate the time in seconds and in hours for a liquid to fall in a tank from a height $z_1 = 9$ m to a height $z_2 = 4$ m above the level of the discharge end of the outlet pipe given the following data:

tank diameter D_T	$= 2$ m
inside diameter of outlet pipe d_i	$= 0.02$ m
dimensionless correction factor α	$= 1$
basic friction factor j_f	$= 0.004$
equivalent length of outlet pipe ΣL_e	$= 25$ m
gravitational acceleration g	$= 9.81$ m/s^2
pressure P_A	$= P_B$

Calculations:

time to empty liquid in tank from height z_1 to height z_2 above the level of the discharge end of the outlet pipe

$$\Delta t (1 - 2) = (z_1^{\frac{1}{2}} - z_2^{\frac{1}{2}})\left(\frac{2D_T^2}{d_i^2}\right)\left[\frac{1/\alpha + 8j_f(\Sigma L_e/d_i)}{2g}\right]^{\frac{1}{2}} \quad (10.1-10)$$

$$z_1^{\frac{1}{2}} = 3 \text{ m}^{\frac{1}{2}}$$

$$z_2^{\frac{1}{2}} = 2 \text{ m}^{\frac{1}{2}}$$

$$D_T = 2 \text{ m}$$

$$d_i = 0.02 \text{ m}$$

$$\frac{2D_T^2}{d_i^2} = 20\,000$$

$$8j_f(\Sigma L_e/d_i) = \frac{(8)(0.004)(25 \text{ m})}{0.02 \text{ m}} = 40$$

$$\frac{1/\alpha + 8j_f(\Sigma L_e/d_i)}{2g} = \frac{41}{(2)(9.81 \text{ m/s}^2)}$$

$$= 2.090 \text{ s}^2/\text{m}$$

$$\left[\frac{1/\alpha + 8j_f(\Sigma L_e/d_i)}{2g}\right]^{\frac{1}{2}} = 1.446 \text{ s/m}^{\frac{1}{2}}$$

$$\Delta t\,(1-2) = (3 \text{ m}^{\frac{1}{2}} - 2 \text{ m}^{\frac{1}{2}})(20\,000)(1.446 \text{ s/m}^{\frac{1}{2}})$$

$$= 28\,920 \text{ s}$$

$$= 8.04 \text{ h}$$

10.2 Time to empty an ideal gas from a tank

Consider the adiabatic flow of an ideal gas from a tank through a constriction in a horizontal conduit to a second tank. The flow rate of gas in mass per unit time is given by equation (6.7–14) for subsonic flow

$$M = S_2\sqrt{\frac{2\gamma P_1\rho_1(P_2/P_1)^{2/\gamma}[1 - (P_2/P_1)^{(\gamma-1)/\gamma}]}{(\gamma - 1)[1 - (S_2/S_1)^2(P_2/P_1)^{2/\gamma}]}} \qquad (6.7\text{–}14)$$

Equation (6.7–14) gives the flow rate of gas M for a given pressure drop driving force $\Delta P = P_1 - P_2$. As the gas flows from the first tank, the pressure P_1 falls and the corresponding flow rate M is reduced. Thus it takes progressively longer for each unit mass of gas to flow out of the tank.

Example (10.2–1)

Nitrogen contained in a 10 m³ tank at a pressure of 200 000 N/m² and a temperature of 300 K flows under adiabatic conditions into a second tank through a horizontal converging nozzle with a 0.05 m diameter throat. The pressure in the second tank and at the nozzle throat is 140 000 N/m². Assume that the mean linear velocity in the nozzle throat is below the critical velocity. Also assume frictionless flow and ideal gas behaviour. Calculate the time in seconds for the pressure in the 10 m³ tank to drop in increments of 1000 N/m² to 190 000 N/m². The following data are given:

gas constant R_G = 8.3143 kJ/(kmol K)
molecular weight of nitrogen (MW) = 28.02 kg/kmol
ratio of heat capacities per unit mass for nitrogen γ = 1.39

Calculations:
flow rate

$$M = S_2 \sqrt{\frac{2\gamma P_1 \rho_1 (P_2/P_1)^{2/\gamma}[1 - (P_2/P_1)^{(\gamma-1)/\gamma}]}{(\gamma - 1)[1 - (S_2/S_1)^2(P_2/P_1)^{2/\gamma}]}} \qquad (6.7\text{--}14)$$

throat diameter $d_2 = 0.05$ m
cross-sectional flow area in throat

$$S_2 = \frac{\pi d_2^2}{4} = 1.964 \times 10^{-3} \text{ m}^2$$

$$\gamma = 1.39$$

$$\gamma - 1 = 0.39$$

$$\frac{\gamma - 1}{\gamma} = 0.2806$$

$$\frac{1}{\gamma} = 0.7194$$

$$P_1 = 200\ 000 \text{ N/m}^2$$

$$P_2 = 140\ 000 \text{ N/m}^2$$

$$\frac{P_2}{P_1} = 0.7000$$

$$\left(\frac{P_2}{P_1}\right)^{2/\gamma} = 0.7000^{1.439} = 0.5985$$

$$\left(\frac{P_2}{P_1}\right)^{(\gamma-1)/\gamma} = 0.7000^{0.281} = 0.9046$$

$$1 - \left(\frac{P_2}{P_1}\right)^{(\gamma-1)/\gamma} = 0.0954$$

since cross-section of tank, S_1, is large, assume $S_2/S_1 \cong 0$; calculate
the density ρ_1 from equation (6.2–4)

$$PV = \frac{R_G T}{(MW)} \qquad (6.2\text{--}4)$$

rewrite equation (6.2–4) in the form

$$\rho_1 = \frac{P_1(MW)}{R_G T_1}$$

$$P_1 = 200\,000 \text{ N/m}^2$$

$$(MW) = 28.02 \text{ kg/kmol}$$

$$R_G = 8.3143 \text{ kJ/(kmol K)}$$

$$T_1 = 300 \text{ K}$$

$$\rho_1 = 2.247 \text{ kg/m}^3$$

flow rate

$$M = (1.964 \times 10^{-3} \text{ m}^2)$$

$$\times \sqrt{\frac{(2)(1.39)(200\,000 \text{ N/m}^2)(2.247 \text{ kg/m}^3)(0.5985)(0.0954)}{0.39}}$$

$$= (1.964 \times 10^{-3} \text{ m}^2)\sqrt{1.829 \times 10^5 \text{ kg}^2/(\text{s}^2 \text{ m}^4)}$$

$$= 0.8399 \text{ kg/s} \quad \text{for} \quad P_1 = 200\,000 \text{ N/m}^2$$

calculate the corresponding flow rates M from equation (6.7–14) for other values of the pressure P_1 and list in Table (10.2–1)
since initially the density

$$\rho_1 = 2.244 \text{ kg/m}^3$$

and the volume of the first tank is 10 m^3, the mass of gas initially in the tank

$$W = 22.44 \text{ kg}$$

the mass of gas in the first tank decreases as the pressure is decreased from $200\,000 \text{ N/m}^2$ to $199\,000 \text{ N/m}^2$ to

$$W = (22.44 \text{ kg})\left(\frac{199\,000 \text{ N/m}^2}{200\,000 \text{ N/m}^2}\right) = 22.328 \text{ kg}$$

$$\Delta W = 22.44 \text{ kg} - 22.328 \text{ kg} = 0.112 \text{ kg}$$

the time taken for this mass of gas to be released at a rate

$$M = 0.8399 \text{ kg/s}$$

TABLE (10.2–1)

P_1 N/m²	ρ_1 kg/m³	P_2/P_1	$(P_2/P_1)^{2/\gamma}$	$(P_2/P_1)^{(\gamma-1)/\gamma}$	$[1-(P_2/P_1)^{(\gamma-1)/\gamma}]$	M kg/s	W kg	ΔW kg	Δt s	$\Sigma\Delta t$ s
200 000	2.247	0.7	0.5985	0.9046	0.0954	0.8399	22.440	0	0	0
199 000	2.235	0.7035	0.6030	0.9059	0.0941	0.8329	22.328	0.112	1.333	1.333
198 000	2.224	0.7071	0.6073	0.9072	0.0928	0.8260	22.216	0.112	1.345	2.678
197 000	2.213	0.7107	0.6117	0.9084	0.0916	0.8197	22.104	0.112	1.356	4.034
196 000	2.202	0.7143	0.6163	0.9097	0.0903	0.8127	21.992	0.112	1.366	5.400
195 000	2.191	0.7179	0.6205	0.9109	0.0891	0.8057	21.880	0.112	1.378	6.778
194 000	2.179	0.7216	0.6253	0.9124	0.0876	0.7978	21.768	0.112	1.390	8.168
193 000	2.168	0.7254	0.6299	0.9137	0.0863	0.7909	21.656	0.112	1.404	9.572
192 000	2.157	0.7292	0.6348	0.9149	0.0851	0.7842	21.544	0.112	1.416	10.988
191 000	2.146	0.7330	0.6396	0.0164	0.0836	0.7763	21.432	0.112	1.428	12.416
190 000	2.134	0.7368	0.6445	0.9177	0.0823	0.7691	21.320	0.112	1.443	13.859

is

$$\Delta t = \frac{\Delta W}{M} = \frac{0.112 \text{ kg}}{0.8399 \text{ kg/s}}$$

$$= 1.333 \text{ s}$$

calculate the corresponding time increments for other increments of pressure loss in a similar manner and list in Table (10.2–1).

10.3 Time to reach 99 per cent of the terminal velocity for a solid spherical particle falling in laminar flow in a Newtonian fluid

Consider a spherical particle of diameter d_p and density ρ_p falling in a fluid of density ρ and dynamic viscosity μ. The particle accelerates until it reaches its terminal settling or falling velocity u_p. At any given time t let the velocity of the particle be u and the acceleration be $\mathrm{d}u/\mathrm{d}t$. A force balance gives

$$\frac{\pi d_p^3}{6}(\rho_p - \rho)g - C_d S_p \frac{\rho u^2}{2} = \frac{\pi d_p^3}{6}(\rho_p - \rho)\frac{\mathrm{d}u}{\mathrm{d}t} \qquad (10.3\text{–}1)$$

where for a spherical particle the projected area $S_p = \pi d_p^2/4$. C_d is a dimensionless drag coefficient and is a function of a Reynolds number defined as

$$(N_{RE})_p = \frac{\rho u d_p}{\mu} \qquad (10.3\text{–}2)$$

For the laminar or streamline Reynolds number range $(N_{RE})_p < 0.2$

$$C_d = \frac{24}{(N_{RE})_p} \qquad (9.1\text{–}6)$$

Substitute equations (9.1–6) and (10.3–2) and $S_p = \pi d_p^2/4$ into equation (10.3–1) to give

$$\frac{\pi d_p^3}{6}(\rho_p - \rho)g - 3\pi d_p \mu u = \frac{\pi d_p^3}{6}(\rho_p - \rho)\frac{\mathrm{d}u}{\mathrm{d}t} \qquad (10.3\text{–}3)$$

which can be simplified to

$$d_p^2(\rho_p - \rho)g - 18\mu u = d_p^2(\rho_p - \rho)\frac{\mathrm{d}u}{\mathrm{d}t} \qquad (10.3\text{–}4)$$

or to

$$\frac{du}{dt} + \frac{18\mu u}{d_p^2(\rho_p - \rho)} = g \tag{10.3-5}$$

The terminal falling or settling velocity in the laminar flow region is given by equation (9.1–7)

$$u_p = \frac{d_p^2(\rho_p - \rho)g}{18\mu} \tag{9.1-7}$$

Combine equations (9.1–7) and (10.3–5) and write

$$\frac{du}{dt} + \left(\frac{g}{u_p}\right)u = g \tag{10.3-6}$$

The solution to equation (10.3–6) can be written in the form

$$u = u_p[1 - e^{-(g/u_p)t}] \tag{10.3-7}$$

Equation (10.3–7) gives the velocity u of a spherical particle at a time t falling in laminar flow under the influence of gravity in terms of the terminal falling or settling velocity u_p.

When $u = 0.99u_p$

$$e^{-(g/u_p)t} = 0.01 \quad \text{and} \quad e^{(g/u_p)t} = 100$$

Therefore the time to reach 99 per cent of the terminal velocity is

$$t = \frac{u_p \ln 100}{g} \tag{10.3-8}$$

Therefore

$$t = 0.469\, u_p$$

where t is the time in seconds and u_p is the terminal falling velocity in m/s.

10.4   Suddenly accelerated plate in a Newtonian liquid[3]

Consider a Newtonian liquid in which molecular momentum transfer occurs in the z direction as shown in Figure (10.4–1). A

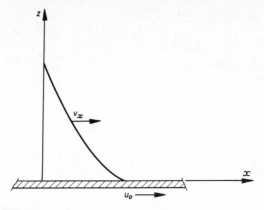

Figure (10.4–1)
Suddenly accelerated plate.

momentum balance for a volume element $(1)(\Delta z)$ can be written as follows:

$$
\left[
\begin{array}{l}
\text{rate of accumulation} \\
\text{of momentum in the} \\
\text{volume element (1) } (\Delta z)
\end{array}
\right]
$$

$$
=
\left[
\begin{array}{l}
\text{rate of transfer} \\
\text{of momentum into} \\
\text{the volume element} \\
\text{(1) } (\Delta z) \text{ by} \\
\text{molecular motion}
\end{array}
\right]
-
\left[
\begin{array}{l}
\text{rate of transfer of} \\
\text{momentum out of the} \\
\text{volume element} \\
\text{(1) } (\Delta z) \text{ by molecular} \\
\text{motion}
\end{array}
\right]
$$

$$
\frac{\partial}{\partial t}(\rho v_x)\,\Delta z = R_{zx}|_z - R_{zx}|_{z+\Delta z} \tag{10.4–1}
$$

Rewrite equation (10.4–1) in the form

$$
\frac{\partial}{\partial t}(\rho v_x)\,\Delta z = \frac{[R_{zx}|_z - R_{zx}|_{z+\Delta z}]\,\Delta z}{\Delta z} \tag{10.4–2}
$$

which in the limit $\Delta z \to 0$ becomes

$$
\frac{\partial}{\partial t}(\rho v_x)\,\Delta z = -\frac{\partial R_{zx}}{\partial z}\Delta z
$$

or

$$
\frac{\partial}{\partial t}(\rho v_x) = -\frac{\partial R_{zx}}{\partial z} \tag{10.4–3}
$$

Newtonian fluids obey equation (1.3–2)

$$R_{zx} = -\mu \frac{dv_x}{dz} \qquad (1.3\text{–}2)$$

Substitute equation (1.3–2) into equation (10.4–3) to give

$$\frac{\partial}{\partial t}(\rho v_x) = \frac{\partial}{\partial z}\left(\mu \frac{\partial v_x}{\partial z}\right) \qquad (10.4\text{–}4)$$

For a Newtonian liquid with a constant density ρ and a constant dynamic viscosity μ, equation (10.4–4) can be written as

$$\frac{\partial v_x}{\partial t} = \eta \frac{\partial^2 v_x}{\partial z^2} \qquad (10.4\text{–}5)$$

where $\eta = \mu/\rho$ is the kinematic viscosity.

Consider a large flat horizontal plate in the x direction above which is a large amount of a stationary Newtonian liquid. At time $t = 0$, the plate is set in motion with a velocity u_o in the x direction. As a result of viscous drag, the Newtonian liquid is also set in motion. The liquid immediately adjacent to the plate is affected most and travels at the same velocity as the plate. The liquid at a large distance away from the plate is not affected at all.

Let the z direction be perpendicular to the plate which lies in the x direction. Consider the fluid flow to be laminar. Determine the point velocity v_x of the Newtonian liquid at any distance z away from the plate at any time t.

Solve equation (10.4–5) written as

$$\frac{\partial v_x}{\partial t} = \eta \frac{d^2 v_x}{dz^2} \qquad (10.4\text{–}6)$$

for the following boundary conditions:

B.C.1 at $t = 0$, $v_x = 0$

B.C.2 at $t > 0$, $v_x = u_0$ at $z = 0$

B.C.3 at $t \geqslant 0$, $v_x = 0$ at $z = \infty$

The first boundary condition implies that initially the plate and the Newtonian liquid are stationary. The second boundary condition implies that the liquid immediately adjacent to the plate travels with the same velocity as the plate. The third boundary condition implies that at an infinite distance away from the plate the liquid remains stationary.

Solve equation (10.4–6) using Laplace transforms[1] where the Laplace transform $\bar{f}$ of a function f is given by the equation

$$\bar{f} = \int_0^\infty f(t)\,e^{-st}\,dt \qquad (10.4\text{–}7)$$

Thus the Laplace transform of the point velocity v_x can be written as

$$\bar{v}_x = \int_0^\infty v_x\,e^{-st}\,dt \qquad (10.4\text{–}8)$$

One property of Laplace transforms is that the time differential of the Laplace transform is related to the Laplace transform by the equation

$$\bar{f}' = s\bar{f} - f(t)|_{t=0} \qquad (10.4\text{–}9)$$

Therefore

$$\bar{v}'_x = s\bar{v}_x - v_x|_{t=0} \qquad (10.4\text{–}10)$$

Combine boundary condition B.C.1 with equation (10.4–10) to give

$$\bar{v}'_x = s\bar{v}_x \qquad (10.4\text{–}11)$$

Write equation (10.4–6) as

$$v'_x = \eta\,\frac{d^2 v_x}{dz^2} \qquad (10.4\text{–}12)$$

and take the Laplace transforms of both sides to give

$$\bar{v}'_x = \eta\,\frac{d^2 \bar{v}_x}{dz^2} \qquad (10.4\text{–}13)$$

Combine equations (10.4–11) and (10.4–13) to give

$$\frac{d^2 \bar{v}_x}{dz^2} = \beta^2 \bar{v}_x \qquad (10.4\text{–}14)$$

where $\beta^2 = s/\eta$.

The solution of equation (10.4–14) is

$$\bar{v}_x = C_1\,e^{\beta z} + C_2\,e^{-\beta z} \qquad (10.4\text{–}15)$$

where C_1 and C_2 are constants.

Combine boundary condition B.C.3 with equation (10.4–15) to give

$$\bar{v}_x = C_2\, e^{-\beta z} \qquad (10.4\text{--}16)$$

Combine boundary condition B.C.2 with equation (10.4–16) to give

$$\bar{u}_o = C_2$$

Therefore

$$C_2 = u_o/s$$

and equation (10.4–16) can be written either as

$$\bar{v}_x = \frac{u_o}{s}\, e^{-\beta z} \qquad (10.4\text{--}17)$$

or as

$$\bar{v}_x = \frac{u_o}{s}\, e^{-z\sqrt{(s/\eta)}} \qquad (10.4\text{--}18)$$

Substitute $k = z/\sqrt{\eta}$ into equation (10.4–18) and write

$$\bar{v}_x = u_o\!\left(\frac{1}{s}\, e^{-k\sqrt{s}}\right) \qquad (10.4\text{--}19)$$

From tables of Laplace transforms[1], read the inverse Laplace transform of $[(1/s)\, e^{-k\sqrt{s}}]$ to be $[1 - \mathrm{erf}\,(k/2\sqrt{t})]$. Take the inverse Laplace transform of each side of equation (10.4–19) and write

$$v_x = u_o\!\left[1 - \mathrm{erf}\!\left(\frac{k}{2\sqrt{t}}\right)\right] \qquad (10.4\text{--}20)$$

Since $k = z/\sqrt{\eta}$, equation (10.4–20) can be written as

$$v_x = u_o\!\left[1 - \mathrm{erf}\!\left(\frac{z}{\sqrt{4\eta t}}\right)\right] \qquad (10.4\text{--}21)$$

The erf function in equations (10.4–20) and (10.4–21) is the well known error function. Tables of it are readily available[2].

Example (10.4–1)

A large flat horizontal plate is immersed in a large volume of Newtonian liquid of density $1000\ \mathrm{kg/m^3}$ and dynamic viscosity

0.1 N s/m². Initially the plate and the liquid are stationary. Suddenly the plate is set in motion with a velocity of 1.0 m/s. Calculate the point linear velocity of the liquid at a distance 0.1 m above the plate at times 25 s and 2500 s after the plate has been set in motion.

Calculations:

$$\rho = 1000 \text{ kg/m}^3$$

$$\mu = 0.1 \text{ N s/m}^2 = 0.1 \text{ kg/(s m)}$$

$$\eta = \mu/\rho = \frac{0.1 \text{ kg/(s m)}}{1000 \text{ kg/m}^3}$$

$$= 1 \times 10^{-4} \text{ m}^2/\text{s}$$

for $t = 25$ s

$$4\eta t = (4)(1 \times 10^{-4} \text{ m}^2/\text{s})(25 \text{ s})$$

$$= 1 \times 10^{-2} \text{ m}^2$$

$$\sqrt{4\eta t} = 1 \times 10^{-1} \text{ m}$$

$$z = 0.1 \text{ m}$$

$$\frac{z}{\sqrt{4\eta t}} = \frac{0.1 \text{ m}}{0.1 \text{ m}} = 1$$

from the tables of error functions[2]

$$\text{erf}(1.0) = 0.8427$$

$$1 - \text{erf}\left(\frac{z}{\sqrt{4\eta t}}\right) = 1 - 0.8427$$

$$= 0.1573$$

$$v_x = u_o\left[1 - \text{erf}\left(\frac{z}{\sqrt{4\eta t}}\right)\right] \qquad (10.4\text{–}21)$$

$$u_o = 1.0 \text{ m/s}$$

$$v_x = (1.0 \text{ m/s})(0.1573)$$

$$= 0.1573 \text{ m/s} \quad \text{at} \quad z = 0.1 \text{ m and } t = 25 \text{ s}$$

for $t = 2500$ s

$$4\eta t = (4)(1 \times 10^{-4} \text{ m}^2/\text{s})(2500 \text{ s})$$

$$= 1.0 \text{ m}^2$$

$$\sqrt{4\eta t} = 1.0 \text{ m}$$

$$z = 0.1 \text{ m}$$

$$\frac{z}{\sqrt{4\eta t}} = \frac{0.1 \text{ m}}{1.0 \text{ m}} = 0.1$$

from the tables of error functions[2]

$$\text{erf}(0.1) = 0.11246$$

$$1 - \text{erf}\left(\frac{z}{\sqrt{4\eta t}}\right) = 1 - 0.11246$$

$$= 0.88754$$

$$v_x = u_o\left[1 - \text{erf}\left(\frac{z}{\sqrt{4\eta t}}\right)\right] \qquad (10.4\text{--}21)$$

$$u_o = 1.0 \text{ m/s}$$

$$v_x = (1.0 \text{ m/s})(0.88754)$$

$$= 0.88754 \text{ m/s} \quad \text{at} \quad z = 0.1 \text{ m and } t = 2500 \text{ s}$$

REFERENCES

(1) Churchill, R. V., *Modern Operational Mathematics in Engineering*, p. 299, New York, McGraw-Hill Book Co. Inc., 1944.
(2) Dale, J. B., *Five-Figure Tables of Mathematical Functions*, 2nd edition, p. 111, London, Edward Arnold (Publishers) Ltd., 1949.
(3) Rohsenow, W. M., and Choi, H. Y., *Heat, Mass, and Momentum Transfer*, p. 46, Englewood Cliffs, New Jersey, Prentice-Hall Inc., 1961.

Part two VECTOR METHODS IN FLUID FLOW

11
Vector methods in fluid flow and the equations of continuity and momentum transfer

11.1 Vectors in fluid flow

For fluid flow in more than one direction it is necessary to consider the components of the velocity vector in various coordinate systems. Scalars such as pressure or temperature have magnitude but no direction. Vectors such as velocity or acceleration have both magnitude and direction. In this text, vectors are written in bold face type. For an x, y, z rectangular Cartesian coordinate system the velocity vector $\mathbf{v}$ can be written in terms of the scalar components v_x, v_y and v_z in the positive, x, y and z directions respectively as

$$\mathbf{v} = \mathbf{i}v_x + \mathbf{j}v_y + \mathbf{k}v_z \tag{11.1-1}$$

where $\mathbf{i}$, $\mathbf{j}$ and $\mathbf{k}$ are defined as unit vectors or vectors of unit magnitude in the positive x, y and z directions respectively. The velocity vector $\mathbf{v}$ is drawn in Figure (11.1–1) with reference to the right handed x, y, z rectangular Cartesian coordinate system.

Any two vectors $\mathbf{a}$ and $\mathbf{b}$ can be written respectively in terms of an x, y, z rectangular Cartesian coordinate system as

$$\mathbf{a} = \mathbf{i}a_x + \mathbf{j}a_y + \mathbf{k}a_z \tag{11.1-2}$$

and

$$\mathbf{b} = \mathbf{i}b_x + \mathbf{j}b_y + \mathbf{k}b_z \tag{11.1-3}$$

Rectangular and cylindrical coordinate systems are the most commonly used in fluid flow. Only the right handed x, y, z rectangular coordinate system is used in this text.

187

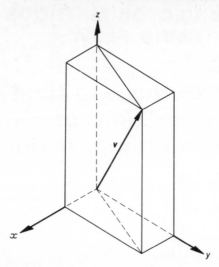

Figure (11.1–1)
Geometric representation of velocity vector **v** in the right handed x, y, z rectangular Cartesian coordinates.

11.2 Scalar product of two vectors in rectangular coordinates

The scalar product of two vectors **a** and **b** is known as the dot or inner product and is defined as

$$\mathbf{a} \cdot \mathbf{b} = ab \cos \theta \qquad (11.2–1)$$

where θ is the angle between the positive directions of the two vectors.

When two vectors **a** and **b** are perpendicular, $\theta = 90°$, $\cos \theta = 0$ and equation (11.2–1) becomes

$$\mathbf{a} \cdot \mathbf{b} = 0 \qquad (11.2–2)$$

The unit vectors **i**, **j** and **k** in x, y, z rectangular coordinates are perpendicular to each other. Thus the following relationships hold:

$$\mathbf{i} \cdot \mathbf{j} = 0 \qquad (11.2–3)$$

$$\mathbf{i} \cdot \mathbf{k} = 0 \qquad (11.2–4)$$

$$\mathbf{j} \cdot \mathbf{k} = 0 \qquad (11.2–5)$$

When two vectors **a** and **b** are parallel, $\theta = 0$, $\cos \theta = 1$ and equation (11.2–1) becomes

$$\mathbf{a} \cdot \mathbf{b} = ab \qquad (11.2–6)$$

Thus the following relationships hold:

$$\mathbf{i} \cdot \mathbf{i} = 1 \qquad (11.2\text{--}7)$$

$$\mathbf{j} \cdot \mathbf{j} = 1 \qquad (11.2\text{--}8)$$

$$\mathbf{k} \cdot \mathbf{k} = 1 \qquad (11.2\text{--}9)$$

The vectors $\mathbf{a}$ and $\mathbf{b}$ can be written respectively as

$$\mathbf{a} = \mathbf{i}a_x + \mathbf{j}a_y + \mathbf{k}a_z \qquad (11.1\text{--}2)$$

$$\mathbf{b} = \mathbf{i}b_x + \mathbf{j}b_y + \mathbf{k}b_z \qquad (11.1\text{--}3)$$

Take the dot product of $\mathbf{a}$ and $\mathbf{b}$ and combine with equations (11.2–3), (11.2–4), (11.2–5), (11.2–7), (11.2–8) and (11.2–9) to give

$$\mathbf{a} \cdot \mathbf{b} = a_x b_x + a_y b_y + a_z b_z \qquad (11.2\text{--}10)$$

which is a scalar quantity.

11.3 Vector product of two vectors in rectangular coordinates

The vector product of two vectors $\mathbf{a}$ and $\mathbf{b}$ is known as the cross or outer product and is defined as

$$\mathbf{a} \times \mathbf{b} = \mathbf{ab} \sin \theta = \mathbf{c} \qquad (11.3\text{--}1)$$

where θ is the angle between the positive directions of the two vectors. The resultant vector $\mathbf{c}$ is perpendicular to the plane of $\mathbf{a}$ and $\mathbf{b}$ as shown in Figure (11.3–1). The direction of $\mathbf{c}$ is governed by the rotation of $\mathbf{a}$ into $\mathbf{b}$ by the right handed rule.

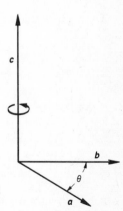

Figure (11.3–1)
Graphical illustration of the vector or cross product of two vectors.

When two vectors **a** and **b** are perpendicular, $\theta = 90°$, $\sin \theta = 1$, and equation (11.3–1) becomes

$$\mathbf{a} \times \mathbf{b} = ab = \mathbf{c} \qquad (11.3\text{–}2)$$

Since the unit vectors **i**, **j** and **k** are perpendicular in the right handed x, y, z rectangular coordinate system the following relationships hold:

$$\mathbf{i} \times \mathbf{j} = \mathbf{k} \qquad (11.3\text{–}3)$$

$$\mathbf{j} \times \mathbf{k} = \mathbf{i} \qquad (11.3\text{–}4)$$

$$\mathbf{k} \times \mathbf{i} = \mathbf{j} \qquad (11.3\text{–}5)$$

$$\mathbf{j} \times \mathbf{i} = -\mathbf{k} \qquad (11.3\text{–}6)$$

$$\mathbf{k} \times \mathbf{j} = -\mathbf{i} \qquad (11.3\text{–}7)$$

$$\mathbf{i} \times \mathbf{k} = -\mathbf{j} \qquad (11.3\text{–}8)$$

When two vectors **a** and **b** are parallel, $\theta = 0$, $\sin \theta = 0$, and equation (11.3–1) becomes

$$\mathbf{a} \times \mathbf{b} = 0 \qquad (11.3\text{–}9)$$

Thus the following relationships hold:

$$\mathbf{i} \times \mathbf{i} = 0 \qquad (11.3\text{–}10)$$

$$\mathbf{j} \times \mathbf{j} = 0 \qquad (11.3\text{–}11)$$

$$\mathbf{k} \times \mathbf{k} = 0 \qquad (11.3\text{–}12)$$

The vectors **a** and **b** can be written respectively as

$$\mathbf{a} = \mathbf{i}a_x + \mathbf{j}a_y + \mathbf{k}a_z \qquad (11.1\text{–}2)$$

$$\mathbf{b} = \mathbf{i}b_x + \mathbf{j}b_y + \mathbf{k}b_z \qquad (11.1\text{–}3)$$

Take the cross product of **a** and **b** and combine with equations (11.3–3), (11.3–4), (11.3–5), (11.3–6), (11.3–7), (11.3–8), (11.3–10), (11.3–11) and (11.3–12) to give

$$\mathbf{a} \times \mathbf{b} = \mathbf{i}(a_y b_z - a_z b_y) - \mathbf{j}(a_x b_z - a_z b_x) + \mathbf{k}(a_x b_y - a_y b_x) \qquad (11.3\text{–}13)$$

which is a vector quantity.

Equation (11.3–13) is also the determinant of the matrix

$$\mathbf{a} \times \mathbf{b} = \begin{vmatrix} \mathbf{i} & \mathbf{j} & \mathbf{k} \\ a_x & a_y & a_z \\ b_x & b_y & b_z \end{vmatrix} \qquad (11.3\text{–}14)$$

11.4   Vector operator del in rectangular coordinates

Del is defined in rectangular coordinates as

$$\mathbf{\nabla} = \mathbf{i}\frac{\partial}{\partial x} + \mathbf{j}\frac{\partial}{\partial y} + \mathbf{k}\frac{\partial}{\partial z} \qquad (11.4\text{–}1)$$

Del operates on a scalar quantity such as pressure P to give

$$\mathbf{\nabla}P = \mathbf{i}\frac{\partial P}{\partial x} + \mathbf{j}\frac{\partial P}{\partial y} + \mathbf{k}\frac{\partial P}{\partial z} \qquad (11.4\text{–}2)$$

$\mathbf{\nabla}P$ is known as grad P.

Del can be multiplied like an ordinary vector. The dot product of del with a vector such as velocity $\mathbf{v}$ gives

$$\mathbf{\nabla} \cdot \mathbf{v} = \frac{\partial v_x}{\partial x} + \frac{\partial v_y}{\partial y} + \frac{\partial v_z}{\partial z} \qquad (11.4\text{–}3)$$

which is similar to equation (11.2–10).

$$\mathbf{a} \cdot \mathbf{b} = a_x b_x + a_y b_y + a_z b_z \qquad (11.2\text{–}10)$$

The dot product of del with the velocity vector $\mathbf{v}$ multiplied by a scalar such as density ρ is

$$\mathbf{\nabla} \cdot (\rho\mathbf{v}) = \frac{\partial}{\partial x}(\rho v_x) + \frac{\partial}{\partial y}(\rho v_y) + \frac{\partial}{\partial z}(\rho v_z) \qquad (11.4\text{–}4)$$

The cross product of del with a vector such as velocity $\mathbf{v}$ gives

$$\mathbf{\nabla} \times \mathbf{v} = \begin{vmatrix} \mathbf{i} & \mathbf{j} & \mathbf{k} \\ \dfrac{\partial}{\partial x} & \dfrac{\partial}{\partial y} & \dfrac{\partial}{\partial z} \\ v_x & v_y & v_z \end{vmatrix} \qquad (11.4\text{–}5)$$

which is similar to equation (11.3–14)

$$\mathbf{a} \times \mathbf{b} = \begin{vmatrix} \mathbf{i} & \mathbf{j} & \mathbf{k} \\ a_x & a_y & a_z \\ b_x & b_y & b_z \end{vmatrix} \qquad (11.3\text{–}14)$$

$\nabla \times \mathbf{v}$ is known as the curl of the velocity vector $\mathbf{v}$.

11.5   Derivation of equation of continuity for fluid flow in rectangular coordinates[2]

A mass balance can be written on a volume element to give

$$\begin{pmatrix} \text{rate of change} \\ \text{of mass} \end{pmatrix} = \begin{pmatrix} \text{rate of accumulation} \\ \text{of mass} \end{pmatrix} \qquad (11.5\text{–}1)$$

and the equation of continuity derived as follows.

Consider a small volume $\Delta x \, \Delta y \, \Delta z$ of fluid in Figure (11.5–1). Let the fluid density be $\rho(x, y, z, t)$ at a particular point (x, y, z) and time t. The components of the velocity vector $\mathbf{v}$ are v_x, v_y and v_z in the positive x, y and z directions respectively. The rate of mass flow into the volume element in the y direction across the surface of area $\Delta x \, \Delta z$ is

$$\Delta x \, \Delta z \, (\rho v_y)|_y$$

The rate of mass flow out of the volume element in the y direction across the surface of area $\Delta x \, \Delta z$ is

$$\Delta x \, \Delta z \, (\rho v_y)|_{y + \Delta y}$$

The rate of accumulation of mass in the volume element in the y direction is therefore

$$\Delta x \, \Delta z \, [(\rho v_y)|_y - (\rho v_y)|_{y + \Delta y}]$$

which can be written as

$$\frac{\Delta x \, \Delta y \, \Delta z \, [(\rho v_y)|_y - (\rho v_y)|_{y + \Delta y}]}{\Delta y}$$

In the limit $\Delta y \to 0$ the accumulation of mass in the volume element in the y direction becomes

$$-\Delta x \, \Delta y \, \Delta z \frac{\partial}{\partial y}(\rho v_y)$$

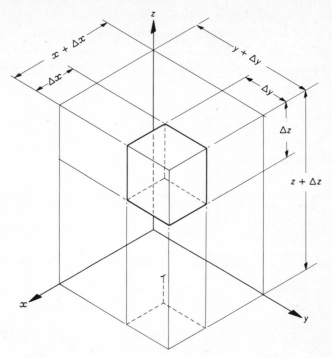

Figure (11.5–1)
A small element of volume in right handed x, y, z rectangular Cartesian coordinates.

Similarly the rates of accumulation of mass in the volume element in the x and z directions are

$$-\Delta x \, \Delta y \, \Delta z \frac{\partial}{\partial x}(\rho v_x)$$

and

$$-\Delta x \, \Delta y \, \Delta z \frac{\partial}{\partial z}(\rho v_z)$$

respectively. Since the total mass in the volume element is $\Delta x \, \Delta y \, \Delta z \, \rho$ its rate of change is

$$\Delta x \, \Delta y \, \Delta z \frac{\partial \rho}{\partial t}$$

This rate of change of mass in the volume element must equal the sum of the rates of accumulation of mass in the x, y and z directions. Therefore

$$\Delta x\,\Delta y\,\Delta z\,\frac{\partial\rho}{\partial t} = -\Delta x\,\Delta y\,\Delta z\,\frac{\partial}{\partial x}(\rho v_x) - \Delta x\,\Delta y\,\Delta z\,\frac{\partial}{\partial y}(\rho v_y)$$

$$- \Delta x\,\Delta y\,\Delta z\,\frac{\partial}{\partial z}(\rho v_z) \qquad (11.5\text{--}2)$$

which can be rewritten as

$$\frac{\partial\rho}{\partial t} + \frac{\partial}{\partial x}(\rho v_x) + \frac{\partial}{\partial y}(\rho v_y) + \frac{\partial}{\partial z}(\rho v_z) = 0 \qquad (11.5\text{--}3)$$

or since

$$\mathbf{V}\cdot(\rho\mathbf{v}) = \frac{\partial}{\partial x}(\rho v_x) + \frac{\partial}{\partial y}(\rho v_y) + \frac{\partial}{\partial z}(\rho v_z) \qquad (11.4\text{--}4)$$

as

$$\frac{\partial\rho}{\partial t} + \mathbf{V}\cdot(\rho\mathbf{v}) = 0 \qquad (11.5\text{--}4)$$

Equation (11.5–4) is true for all types of fluids whether compressible or incompressible, Newtonian or non-Newtonian.

11.6   Substantial derivative and alternative forms of the equation of continuity

Carry out the differentiation in equation (11.5–3) to give

$$\frac{\partial\rho}{\partial t} + v_x\frac{\partial\rho}{\partial x} + v_y\frac{\partial\rho}{\partial y} + v_z\frac{\partial\rho}{\partial z} + \rho\left(\frac{\partial v_x}{\partial x} + \frac{\partial v_y}{\partial y} + \frac{\partial v_z}{\partial z}\right) = 0 \quad (11.6\text{--}1)$$

Since

$$\mathbf{V}\cdot\mathbf{v} = \frac{\partial v_x}{\partial x} + \frac{\partial v_y}{\partial y} + \frac{\partial v_z}{\partial z} \qquad (11.4\text{--}3)$$

equation (11.6–1) can be written as

$$\frac{\partial\rho}{\partial t} + v_x\frac{\partial\rho}{\partial x} + v_y\frac{\partial\rho}{\partial y} + v_z\frac{\partial\rho}{\partial z} + \rho\mathbf{V}\cdot\mathbf{v} = 0 \qquad (11.6\text{--}2)$$

The substantial derivative is a vector operator defined as

$$\frac{D}{Dt} = \frac{\partial}{\partial t} + v_x \frac{\partial}{\partial x} + v_y \frac{\partial}{\partial y} + v_z \frac{\partial}{\partial z} \qquad (11.6\text{–}3)$$

Combine equations (11.6–2) and (11.6–3) to give

$$\frac{D\rho}{Dt} + \rho \mathbf{V} \cdot \mathbf{v} = 0 \qquad (11.6\text{–}4)$$

Equation (11.6–4) is the equation of continuity in terms of the substantial derivative. It is true for all types of fluids whether compressible or incompressible, Newtonian or non-Newtonian. Equation (11.5–4) is written from the viewpoint of a stationary observer. Equation (11.6–4) is written from the viewpoint of an observer moving at the velocity of the fluid stream $\mathbf{v}$.

Consider an incompressible fluid where the density ρ is not a function of the coordinates x, y, z or of the time t. For this case equation (11.5–3) can be written either as

$$\frac{\partial v_x}{\partial x} + \frac{\partial v_y}{\partial y} + \frac{\partial v_z}{\partial z} = 0 \qquad (11.6\text{–}5)$$

or in terms of equation (11.4–3) as

$$\mathbf{V} \cdot \mathbf{v} = 0 \qquad (11.6\text{–}6)$$

11.7 Laplacian operator in rectangular coordinates

The Laplacian operator is defined as

$$\mathbf{V}^2 = \frac{\partial^2}{\partial x^2} + \frac{\partial^2}{\partial y^2} + \frac{\partial^2}{\partial z^2} \qquad (11.7\text{–}1)$$

It operates on a scalar quantity such as the velocity component v_x in the x direction of a rectangular coordinate system to give

$$\mathbf{V}^2 v_x = \frac{\partial^2 v_x}{\partial x^2} + \frac{\partial^2 v_x}{\partial y^2} + \frac{\partial^2 v_x}{\partial z^2} \qquad (11.7\text{–}2)$$

The corresponding equations for the velocity components v_y and v_z in the y and z directions respectively are

$$\mathbf{V}^2 v_y = \frac{\partial^2 v_y}{\partial x^2} + \frac{\partial^2 v_y}{\partial y^2} + \frac{\partial^2 v_y}{\partial z^2} \qquad (11.7\text{–}3)$$

and

$$\nabla^2 v_z = \frac{\partial^2 v_z}{\partial x^2} + \frac{\partial^2 v_z}{\partial y^2} + \frac{\partial^2 v_z}{\partial z^2} \tag{11.7-4}$$

Since

$$\mathbf{v} = \mathbf{i}v_x + \mathbf{j}v_y + \mathbf{k}v_z \tag{11.1-1}$$

equations (11.7–2), (11.7–3) and (11.7–4) can be combined to give

$$\nabla^2 \mathbf{v} = \mathbf{i}\left(\frac{\partial^2 v_x}{\partial x^2} + \frac{\partial^2 v_x}{\partial y^2} + \frac{\partial^2 v_x}{\partial z^2}\right)$$

$$+ \mathbf{j}\left(\frac{\partial^2 v_y}{\partial x^2} + \frac{\partial^2 v_y}{\partial y^2} + \frac{\partial^2 v_y}{\partial z^2}\right)$$

$$+ \mathbf{k}\left(\frac{\partial^2 v_z}{\partial x^2} + \frac{\partial^2 v_z}{\partial y^2} + \frac{\partial^2 v_z}{\partial z^2}\right) \tag{11.7-5}$$

11.8   Cylindrical coordinates

Many fluid flow systems are more conveniently represented in cylindrical coordinates than in rectangular coordinates. Both the r, θ, x horizontal cylindrical coordinate system and the r, θ, z vertical cylindrical coordinate system are widely used where r is the radial distance from either the x or z axis and θ is the angle the radius r makes with a fixed line which is perpendicular to the x or z axis.

The r, θ, x horizontal cylindrical coordinate system is related to the x, y, z rectangular coordinate system as follows:

$$x = r \cos \theta \tag{11.8-1}$$

$$y = r \sin \theta \tag{11.8-2}$$

$$z = x \tag{11.8-3}$$

The r, θ, z vertical cylindrical coordinate system is related to the x, y, z rectangular coordinate system as follows:

$$x = r \cos \theta \tag{11.8-4}$$

$$y = r \sin \theta \tag{11.8-5}$$

$$z = z \tag{11.8-6}$$

Equations (11.8–1), (11.8–2), (11.8–3), (11.8–4), (11.8–5) and (11.8–6) can be used to transform the various vector operators from rectangular to cylindrical coordinates[1,3].

It can readily be shown[3] that the vector operator del can be written in vertical cylindrical coordinates as

$$\mathbf{V} = \mathbf{i}_r\frac{\partial}{\partial r} + \mathbf{i}_\theta\frac{1}{r}\frac{\partial}{\partial\theta} + \mathbf{i}_z\frac{\partial}{\partial z} \qquad (11.8\text{--}7)$$

where $\mathbf{i}_r$, $\mathbf{i}_\theta$ and $\mathbf{i}_z$ are unit vectors in the r, θ and z directions respectively. The dot product of del with a vector such as velocity $\mathbf{v}$ in vertical cylindrical coordinates gives

$$\mathbf{V}\cdot\mathbf{v} = \frac{1}{r}\frac{\partial}{\partial r}(rv_r) + \frac{1}{r}\frac{\partial v_\theta}{\partial\theta} + \frac{\partial v_z}{\partial z} \qquad (11.8\text{--}8)$$

The cross product of del with a vector such as velocity $\mathbf{v}$ in vertical cylindrical coordinates gives

$$\mathbf{V}\times\mathbf{v} = \frac{1}{r}\begin{vmatrix} \mathbf{i}_r & \mathbf{i}_\theta & \mathbf{i}_z \\ \dfrac{\partial}{\partial r} & \dfrac{\partial}{\partial\theta} & \dfrac{\partial}{\partial z} \\ v_r & rv_\theta & v_z \end{vmatrix} \qquad (11.8\text{--}9)$$

It can also be readily shown[1] that the Laplacian operator can be written in vertical cylindrical coordinates as

$$\mathbf{V}^2 = \frac{1}{r}\frac{\partial}{\partial r}\left(r\frac{\partial}{\partial r}\right) + \frac{1}{r^2}\frac{\partial^2}{\partial\theta^2} + \frac{\partial^2}{\partial z^2} \qquad (11.8\text{--}10)$$

11.9 Equation of continuity for fluid flow in cylindrical coordinates

The equation of continuity is

$$\frac{\partial\rho}{\partial t} + \mathbf{V}\cdot(\rho\mathbf{v}) = 0 \qquad (11.5\text{--}4)$$

In horizontal cylindrical coordinates it can be written as

$$\frac{\partial\rho}{\partial t} + \frac{1}{r}\frac{\partial}{\partial r}(\rho rv_r) + \frac{1}{r}\frac{\partial}{\partial\theta}(\rho v_\theta) + \frac{\partial}{\partial x}(\rho v_x) = 0 \qquad (11.9\text{--}1)$$

In vertical cylindrical coordinates it can be written as

$$\frac{\partial\rho}{\partial t} + \frac{1}{r}\frac{\partial}{\partial r}(\rho rv_r) + \frac{1}{r}\frac{\partial}{\partial\theta}(\rho v_\theta) + \frac{\partial}{\partial z}(\rho v_z) = 0 \qquad (11.9\text{--}2)$$

11.10   Derivation of general equations for momentum transfer in rectangular coordinates[4]

A momentum balance can be written on a volume element to give

$$
\begin{pmatrix} \text{rate of} \\ \text{change of} \\ \text{momentum} \end{pmatrix} = \begin{pmatrix} \text{rate of} \\ \text{accumulation} \\ \text{of momentum} \\ \text{by convection,} \\ \text{i.e. by bulk} \\ \text{fluid flow} \end{pmatrix} + \begin{pmatrix} \text{rate of} \\ \text{accumulation} \\ \text{of momentum} \\ \text{by molecular} \\ \text{transfer, i.e.} \\ \text{by velocity} \\ \text{gradients} \end{pmatrix} + \begin{pmatrix} \text{sum of} \\ \text{forces} \\ \text{acting} \\ \text{on the} \\ \text{system} \end{pmatrix}
$$

$$(11.10\text{--}1)$$

and the momentum conservation equations derived in right handed x, y, z rectangular Cartesian coordinates as follows.

Consider a small volume $\Delta x \, \Delta y \, \Delta z$ of fluid in Figure (11.5–1). Let the fluid density be $\rho(x, y, z, t)$ at a particular point (x, y, z) and time t. The components of the velocity vector $\mathbf{v}$ are v_x, v_y and v_z in the positive x, y and z directions respectively. Let R_{xx}, R_{yx}, R_{zx} be the fluxes of the x component of momentum, resulting from molecular transfer, through the faces perpendicular to the x, y and z directions, respectively.

The rate of flow of the x component of convection momentum into the volume element in the y direction across the surface of area $\Delta x \, \Delta z$ is

$$\Delta x \, \Delta z \, (\rho v_x v_y)|_y$$

The rate of flow of the x component of convection momentum out of the volume element in the y direction across the surface of area $\Delta x \, \Delta z$ is

$$\Delta x \, \Delta z \, (\rho v_x v_y)|_{y+\Delta y}$$

The rate of accumulation of the x component of convection momentum in the y direction is therefore

$$\Delta x \, \Delta z \, [(\rho v_x v_y)|_y - (\rho v_x v_y)|_{y+\Delta y}]$$

which can be written as

$$\frac{\Delta x \, \Delta y \, \Delta z \, [(\rho v_x v_y)|_y - (\rho v_x v_y)|_{y+\Delta y}]}{\Delta y}$$

In the limit $\Delta y \to 0$ the accumulation of the x component of convection momentum in the y direction becomes

$$- \Delta x \, \Delta y \, \Delta z \frac{\partial}{\partial y}(\rho v_x v_y)$$

Similarly, the rates of accumulation of the x component of convection momentum in the x and z directions are

$$- \Delta x \, \Delta y \, \Delta z \frac{\partial}{\partial x}(\rho v_x v_x)$$

and

$$- \Delta x \, \Delta y \, \Delta z \frac{\partial}{\partial z}(\rho v_x v_z)$$

respectively.

The rate of flow of the x component of molecular transfer momentum into the volume element in the y direction across the surface of area $\Delta x \, \Delta z$ is

$$\Delta x \, \Delta z \, R_{yx}|_y$$

The rate of flow of the x component of molecular transfer momentum out of the volume element in the y direction across the surface of area $\Delta x \, \Delta z$ is

$$\Delta x \, \Delta z \, R_{yx}|_{y+\Delta y}$$

The rate of accumulation of the x component of molecular transfer momentum in the y direction is therefore

$$\Delta x \, \Delta z \, (R_{yx}|_y - R_{yx}|_{y+\Delta y})$$

which can be written as

$$\Delta x \, \Delta y \, \Delta z \frac{(R_{yx}|_y - R_{yx}|_{y+\Delta y})}{\Delta y}$$

In the limit $\Delta y \to 0$ the accumulation of the x component of molecular transfer momentum in the y direction becomes

$$- \Delta x \, \Delta y \, \Delta z \frac{\partial R_{yx}}{\partial y}$$

Similarly, the rates of accumulation of the x component of molecular transfer momentum in the x and z directions are

$$-\Delta x\,\Delta y\,\Delta z\,\frac{\partial R_{xx}}{\partial x} \quad \text{and} \quad -\Delta x\,\Delta y\,\Delta z\,\frac{\partial R_{zx}}{\partial z}$$

respectively.

The pressure and gravitational forces on the volume element in the y direction can be written as

$$\Delta x\,\Delta z\,(P|_y - P|_{y+\Delta y}) + \Delta x\,\Delta y\,\Delta z\,\rho g_y$$

which can also be written as

$$\Delta x\,\Delta y\,\Delta z\,\frac{(P|_y - P|_{y+\Delta y})}{\Delta y} + \Delta x\,\Delta y\,\Delta z\,\rho g_y$$

In the limit $\Delta y \rightarrow 0$ the pressure and gravitational forces on the volume element in the y direction become

$$-\Delta x\,\Delta y\,\Delta z\,\frac{\partial P}{\partial y} + \Delta x\,\Delta y\,\Delta z\,\rho g_y$$

Similarly, the pressure and gravitational forces on the volume element in the x and z directions are

$$-\Delta x\,\Delta y\,\Delta z\,\frac{\partial P}{\partial x} + \Delta x\,\Delta y\,\Delta z\,\rho g_x$$

and

$$-\Delta x\,\Delta y\,\Delta z\,\frac{\partial P}{\partial z} + \Delta x\,\Delta y\,\Delta z\,\rho g_z$$

respectively.

The rate of accumulation of the x component of momentum in the volume element is

$$\Delta x\,\Delta y\,\Delta z\,\frac{\partial}{\partial t}(\rho v_x)$$

Therefore a balance on the x component of momentum can be written for unit volume of fluid as follows:

$$\frac{\partial}{\partial t}(\rho v_x) = -\left[\frac{\partial}{\partial x}(\rho v_x v_x) + \frac{\partial}{\partial y}(\rho v_x v_y) + \frac{\partial}{\partial z}(\rho v_x v_z)\right]$$

$$-\left(\frac{\partial R_{xx}}{\partial x} + \frac{\partial R_{yx}}{\partial y} + \frac{\partial R_{zx}}{\partial z}\right) - \frac{\partial P}{\partial x} + \rho g_x \quad (11.10\text{–}2)$$

Differentiate equation (11.10–2) to give

$$\rho \left(\frac{\partial v_x}{\partial t} + v_x \frac{\partial v_x}{\partial x} + v_y \frac{\partial v_x}{\partial y} + v_z \frac{\partial v_x}{\partial z} \right)$$

$$= -v_x \left[\frac{\partial \rho}{\partial t} + \frac{\partial}{\partial x}(\rho v_x) + \frac{\partial}{\partial y}(\rho v_y) + \frac{\partial}{\partial z}(\rho v_z) \right]$$

$$- \left(\frac{\partial R_{xx}}{\partial x} + \frac{\partial R_{yx}}{\partial y} + \frac{\partial R_{zx}}{\partial z} \right) - \frac{\partial P}{\partial x} + \rho g_x \qquad (11.10\text{–}3)$$

The first term on the right hand side of equation (11.10–3) is zero from equation (11.5–3), the equation of continuity.

$$\frac{\partial \rho}{\partial t} + \frac{\partial}{\partial x}(\rho v_x) + \frac{\partial}{\partial y}(\rho v_y) + \frac{\partial}{\partial z}(\rho v_z) = 0 \qquad (11.5\text{–}3)$$

Since the substantial derivative

$$\frac{D}{Dt} = \frac{\partial}{\partial t} + v_x \frac{\partial}{\partial x} + v_y \frac{\partial}{\partial y} + v_z \frac{\partial}{\partial z} \qquad (11.6\text{–}3)$$

equation (11.10–3), the equation for momentum transfer for the x direction, can be written as

$$\rho \frac{Dv_x}{Dt} = -\left(\frac{\partial R_{xx}}{\partial x} + \frac{\partial R_{yx}}{\partial y} + \frac{\partial R_{zx}}{\partial z} \right) - \frac{\partial P}{\partial x} + \rho g_x \qquad (11.10\text{–}4)$$

The corresponding equations for momentum transfer for the y and z directions are

$$\rho \frac{Dv_y}{Dt} = -\left(\frac{\partial R_{xy}}{\partial x} + \frac{\partial R_{yy}}{\partial y} + \frac{\partial R_{zy}}{\partial z} \right) - \frac{\partial P}{\partial y} + \rho g_y \qquad (11.10\text{–}5)$$

and

$$\rho \frac{Dv_z}{Dt} = -\left(\frac{\partial R_{xz}}{\partial x} + \frac{\partial R_{yz}}{\partial y} + \frac{\partial R_{zz}}{\partial z} \right) - \frac{\partial P}{\partial z} + \rho g_z \qquad (11.10\text{–}6)$$

respectively.

Equations (11.10–4), (11.10–5) and (11.10–6) are true for all fluids whether compressible or incompressible, Newtonian or non-Newtonian.

REFERENCES

(1) Hildebrand, F. B., *Advanced Calculus for Applications*, p. 302, Englewood Cliffs, New Jersey, Prentice-Hall Inc., 1962.
(2) Holland, F. A., and Chapman, F. S., *Pumping of Liquids*, p. 10, New York, Reinhold Publishing Corporation, 1966.
(3) Ibid., p. 18.
(4) Holland, F. A., Moores, R. M., Watson, F. A., and Wilkinson, J. K., *Heat Transfer*, p. 428, London, Heinemann Educational Books Ltd., 1970.

12
Applications of modified Navier Stokes equations in rectangular coordinates

12.1 The modified Navier Stokes equations in rectangular coordinates

The general equation for momentum transfer for the x direction can be written as

$$\rho\frac{Dv_x}{Dt} = -\left(\frac{\partial R_{xx}}{\partial x} + \frac{\partial R_{yx}}{\partial y} + \frac{\partial R_{zx}}{\partial z}\right) - \frac{\partial P}{\partial x} + \rho g_x \quad (11.10\text{--}4)$$

where R_{xx}, R_{yx} and R_{zx} are the fluxes of the x component of momentum, resulting from molecular transfer, through faces perpendicular to x, y and z directions, respectively.

For incompressible fluids of constant dynamic viscosity it can be shown[5,6] that equation (11.10–4) becomes

$$\rho\frac{Dv_x}{Dt} = \mu\left(\frac{\partial^2 v_x}{\partial x^2} + \frac{\partial^2 v_x}{\partial y^2} + \frac{\partial^2 v_x}{\partial z^2}\right) - \frac{\partial P}{\partial x} + \rho g_x \quad (12.1\text{--}1)$$

Similarly for incompressible fluids of constant dynamic viscosity it can be shown[5,6] that equations (11.10–5) and (11.10–6) for momentum transfer in the y and z directions can be written respectively as follows:

$$\rho\frac{Dv_y}{Dt} = \mu\left(\frac{\partial^2 v_y}{\partial x^2} + \frac{\partial^2 v_y}{\partial y^2} + \frac{\partial^2 v_y}{\partial z^2}\right) - \frac{\partial P}{\partial y} + \rho g_y \quad (12.1\text{--}2)$$

and

$$\rho \frac{Dv_z}{Dt} = \mu \left(\frac{\partial^2 v_z}{\partial x^2} + \frac{\partial^2 v_z}{\partial y^2} + \frac{\partial^2 v_z}{\partial z^2} \right) - \frac{\partial P}{\partial z} + \rho g_z \qquad (12.1\text{-}3)$$

The transformation of equations (11.10–4), (11.10–5) and (11.10–6) to equations (12.1–1), (12.1–2) and (12.1–3) respectively is fairly lengthy and of limited interest. The details are readily available in a number of texts.[1,3,5,6]

The velocity vector **v** in x, y, z rectangular coordinates can be written as

$$\mathbf{v} = \mathbf{i}v_x + \mathbf{j}v_y + \mathbf{k}v_z \qquad (11.1\text{-}1)$$

The gravitational acceleration vector **g** in x, y, z rectangular coordinates can be written as

$$\mathbf{g} = \mathbf{i}g_x + \mathbf{j}g_y + \mathbf{k}g_z \qquad (12.1\text{-}4)$$

The product of del and pressure P is

$$\nabla P = \mathbf{i}\frac{\partial P}{\partial x} + \mathbf{j}\frac{\partial P}{\partial y} + \mathbf{k}\frac{\partial P}{\partial z} \qquad (11.4\text{-}2)$$

The product of the Laplacian operator and the velocity vector **v** is

$$\nabla^2 \mathbf{v} = \mathbf{i} \left(\frac{\partial^2 v_x}{\partial x^2} + \frac{\partial^2 v_x}{\partial y^2} + \frac{\partial^2 v_x}{\partial z^2} \right) + \mathbf{j} \left(\frac{\partial^2 v_y}{\partial x^2} + \frac{\partial^2 v_y}{\partial y^2} + \frac{\partial^2 v_y}{\partial z^2} \right)$$

$$+ \mathbf{k} \left(\frac{\partial^2 v_z}{\partial x^2} + \frac{\partial^2 v_z}{\partial y^2} + \frac{\partial^2 v_z}{\partial z^2} \right) \qquad (11.7\text{-}5)$$

Combine equations (12.1–1), (12.1–2), (12.1–3), (11.1–1), (12.1–4), (11.4–2) and (11.7–5) to give

$$\rho \frac{D\mathbf{v}}{Dt} = \mu \nabla^2 \mathbf{v} - \nabla P + \rho \mathbf{g} \qquad (12.1\text{-}5)$$

Equations (12.1–1), (12.1–2), (12.1–3) and (12.1–5) are the modified Navier Stokes equations for incompressible fluids of constant dynamic viscosity. Thus the modified Navier Stokes equations hold for Newtonian liquids of constant density and dynamic viscosity.

Equations (12.1–1), (12.1–2) and (12.1–3) are commonly used in their expanded forms which are written respectively as follows:

$$\rho\left(\frac{\partial v_x}{\partial t} + v_x\frac{\partial v_x}{\partial x} + v_y\frac{\partial v_x}{\partial y} + v_z\frac{\partial v_x}{\partial z}\right)$$

$$= -\frac{\partial P}{\partial x} + \mu\left(\frac{\partial^2 v_x}{\partial x^2} + \frac{\partial^2 v_x}{\partial y^2} + \frac{\partial^2 v_x}{\partial z^2}\right) + \rho g_x \quad (12.1\text{–}6)$$

$$\rho\left(\frac{\partial v_y}{\partial t} + v_x\frac{\partial v_y}{\partial x} + v_y\frac{\partial v_y}{\partial y} + v_z\frac{\partial v_y}{\partial z}\right)$$

$$= -\frac{\partial P}{\partial y} + \mu\left(\frac{\partial^2 v_y}{\partial x^2} + \frac{\partial^2 v_y}{\partial y^2} + \frac{\partial^2 v_y}{\partial z^2}\right) + \rho g_y \quad (12.1\text{–}7)$$

and

$$\rho\left(\frac{\partial v_z}{\partial t} + v_x\frac{\partial v_z}{\partial x} + v_y\frac{\partial v_z}{\partial y} + v_z\frac{\partial v_z}{\partial z}\right)$$

$$= -\frac{\partial P}{\partial z} + \mu\left(\frac{\partial^2 v_z}{\partial x^2} + \frac{\partial^2 v_z}{\partial y^2} + \frac{\partial^2 v_z}{\partial z^2}\right) + \rho g_z \quad (12.1\text{–}8)$$

When viscous effects are insignificant, equation (12.1–5) can be written as

$$\rho\frac{D\mathbf{v}}{Dt} = -\nabla P + \rho\mathbf{g} \quad (12.1\text{–}9)$$

Equation (12.1–9) is the Euler equation for momentum transfer in ideal fluids.

12.2   Steady horizontal laminar flow of a Newtonian liquid

Let x and z be the horizontal and vertical upward directions respectively. Let there be no pressure gradient in the y direction and no fluid flow in the y and z directions.

The following relationships hold:

for gravitational acceleration

$$g_x = 0, \qquad g_y = 0, \qquad g_z = -g$$

for velocity

$$v_y = 0, \qquad v_z = 0, \qquad v_x = v_x(z),$$

for steady flow

$$\frac{\partial v_x}{\partial t} = 0, \qquad \frac{\partial v_x}{\partial x} = 0, \qquad \frac{\partial v_x}{\partial y} = 0,$$

for pressure gradient

$$\frac{\partial P}{\partial y} = 0.$$

For these conditions the expanded modified Navier Stokes equations, equations (12.1–6), (12.1–7) and (12.1–8) reduce to

$$0 = -\frac{\partial P}{\partial x} + \mu \frac{\partial^2 v_x}{\partial z^2} \tag{12.2–1}$$

and

$$0 = -\frac{\partial P}{\partial z} - \rho g \tag{12.2–2}$$

Integrate equation (12.2–1) to

$$\frac{dv_x}{dz} = \frac{1}{\mu}\left(\frac{dP}{dx}\right)z + C_1 \tag{12.2–3}$$

and then to

$$v_x = \frac{1}{\mu}\left(\frac{dP}{dx}\right)\frac{z^2}{2} + C_1 z + C_2 \tag{12.2–4}$$

where C_1 and C_2 are constants.

Equation (12.2–4) gives the point linear velocity v_x in the horizontal direction in terms of position z in the vertical z direction for a Newtonian liquid of constant density and dynamic viscosity.

12.3   Steady horizontal laminar flow of a Newtonian liquid between two infinitely large parallel plates

Let the two infinitely large parallel plates be at distance b apart. Equation (12.2–4) gives the point linear velocity v_x in the horizontal x direction in terms of position z in the vertical z direction for a Newtonian liquid of constant density ρ and dynamic viscosity μ.

$$v_x = \frac{1}{\mu}\left(\frac{dP}{dx}\right)\frac{z^2}{2} + C_1 z + C_2 \tag{12.2–4}$$

Solve equation (12.2–4) for the following boundary conditions:

$$\text{B.C.1 at } z = 0, \qquad v_x = 0$$

$$\text{B.C.2 at } z = b, \qquad v_x = 0$$

The boundary conditions assume that the linear velocity is zero at the lower and upper solid surfaces.

From the boundary condition B.C.1 the constant $C_2 = 0$. From the boundary condition B.C.2 the constant

$$C_1 = -\frac{1}{\mu}\left(\frac{dP}{dx}\right)\frac{b}{2} \qquad (12.3\text{–}1)$$

Substitute equation (12.3–1) into equation (12.2–4) to give

$$v_x = \frac{1}{\mu}\left(\frac{dP}{dx}\right)\left(\frac{z^2}{2} - \frac{bz}{2}\right) \qquad (12.3\text{–}2)$$

or

$$v_x = \frac{z}{2\mu}\left(\frac{dP}{dx}\right)(z - b) \qquad (12.3\text{–}3)$$

Equation (12.3–3) gives the point linear velocity v_x in the horizontal x direction at any vertical position z in between the two parallel plates for a Newtonian liquid of constant density and dynamic viscosity.

The mean linear velocity u in the horizontal x direction is given by the equation

$$u = \frac{1}{b}\int_0^b v_x\,dx \qquad (12.3\text{–}4)$$

Substitute equation (12.3–3) into equation (12.3–4) and carry out the integration to give

$$u = -\frac{b^2}{12\mu}\left(\frac{dP}{dx}\right) \qquad (12.3\text{–}5)$$

The velocity u is positive since the velocity gradient dP/dx is negative.

12.4 Steady boundary layer flow of a Newtonian liquid over a horizontal flat plate[2,4]

In the region of immediate proximity to a solid surface, the fluid is directly affected by the solid surface. This region is known as the boundary layer.

Let x and z be the horizontal and vertical upward directions respectively. Let there be no pressure gradient and fluid flow in the y direction.

The following relationships hold:
for gravitational acceleration

$$g_x = 0, \qquad g_y = 0, \qquad g_z = -g$$

for velocity

$$v_y = 0, \qquad v_x = v_x(z)$$

for steady flow

$$\frac{\partial v_x}{\partial t} = 0, \qquad \frac{\partial v_z}{\partial t} = 0$$

$$\frac{\partial v_x}{\partial y} = 0, \qquad \frac{\partial v_z}{\partial y} = 0$$

for pressure gradient

$$\frac{\partial P}{\partial y} = 0$$

For these conditions the expanded modified Navier Stokes equations, equations (12.1–6), (12.1–7) and (12.1–8) reduce to

$$\rho\left(v_x\frac{\partial v_x}{\partial x} + v_z\frac{\partial v_x}{\partial z}\right) = -\frac{\partial P}{\partial x} + \mu\left(\frac{\partial^2 v_x}{\partial x^2} + \frac{\partial^2 v_x}{\partial z^2}\right) \qquad (12.4\text{–}1)$$

and

$$\rho\left(v_x\frac{\partial v_z}{\partial x} + v_z\frac{\partial v_z}{\partial z}\right) = -\frac{\partial P}{\partial z} + \mu\left(\frac{\partial^2 v_z}{\partial x^2} + \frac{\partial^2 v_z}{\partial z^2}\right) - \rho g \qquad (12.4\text{–}2)$$

The pressure gradient $\partial P/\partial x$ is very small and can be neglected. The vertical velocity v_z is also very small and the differentials can be neglected. In addition the term $\partial^2 v_x/\partial x^2$ is sufficiently small to be neglected. For these conditions equations (12.4–1) and (12.4–2) can be written respectively as

$$v_x\frac{\partial v_x}{\partial x} + v_z\frac{\partial v_x}{\partial z} = \eta\frac{\partial^2 v_x}{\partial z^2} \qquad (12.4\text{–}3)$$

and

$$0 = -\frac{\partial P}{\partial z} - \rho g \qquad (12.4\text{–}4)$$

where the kinematic viscosity $\eta = \mu/\rho$.

The equation of continuity is

$$\frac{\partial \rho}{\partial t} + \frac{\partial}{\partial x}(\rho v_x) + \frac{\partial}{\partial y}(\rho v_y) + \frac{\partial}{\partial z}(\rho v_z) = 0 \qquad (11.5\text{–}3)$$

For this case where the flow is steady, the density is constant and $v_y = 0$, equation (11.5–3) becomes

$$\frac{\partial v_x}{\partial x} + \frac{\partial v_z}{\partial z} = 0 \qquad (12.4\text{–}5)$$

or

$$\frac{\partial v_z}{\partial z} = -\frac{\partial v_x}{\partial x} \qquad (12.4\text{–}6)$$

Let the Newtonian liquid approach the flat plate in the x direction with a uniform stream velocity u_s. At the point of impact with the flat plate, the thickness of the boundary layer is zero. The thickness of the boundary layer increases with the distance x along the flat plate. At any point x, let the thickness of the boundary layer be δ. Solve equation (12.4–3) for the following boundary conditions

$$\text{B.C.1 at } z = 0, \qquad v_x = 0$$
$$\text{B.C.2 at } z = \delta, \qquad v_x = u_s$$

The first boundary condition assumes that the velocity is zero at the surface of the plate. The second boundary condition assumes that at the upper surface of the boundary layer the velocity of the fluid in the x direction is equal to that of the stream velocity u_s.

Equation (12.4–3) can be written for integration over the boundary layer as

$$\int_0^\delta v_x \frac{\partial v_x}{\partial x}\, dz + \int_0^\delta v_z \frac{\partial v_x}{\partial z}\, dz = \int_0^\delta \eta \frac{\partial^2 v_x}{\partial z^2}\, dz \qquad (12.4\text{–}7)$$

Since

$$v_z v_x = \int v_z\, dv_x + \int v_x\, dv_z \qquad (12.4\text{–}8)$$

and

$$v_z v_x = \int v_z \frac{\partial v_x}{\partial z}\, dz + \int v_x \frac{\partial v_z}{\partial z}\, dz \qquad (12.4\text{–}9)$$

equation (12.4–7) can be written in the form

$$\int_0^\delta v_x \frac{\partial v_x}{\partial x}\,dz + (v_z v_x)_0^\delta - \int_0^\delta v_x \frac{\partial v_z}{\partial z}\,dz = \int_0^\delta \eta \frac{\partial^2 v_x}{\partial z^2}\,dz$$

(12.4–10)

Substitute equation (12.4–6) into equation (12.4–10) to give

$$\int_0^\delta 2 v_x \frac{\partial v_x}{\partial x}\,dz + (v_z v_x)_0^\delta = \int_0^\delta \eta \frac{\partial^2 v_x}{\partial z^2}\,dz \qquad (12.4–11)$$

or

$$\int_0^\delta \frac{\partial v_x^2}{\partial x}\,dz + (v_z v_x)_0^\delta = \int_0^\delta \eta \frac{\partial^2 v_x}{\partial z^2}\,dz \qquad (12.4–12)$$

Integrate equation (12.4–6) to give

$$(v_z)_0^\delta = -\int_0^\delta \frac{\partial v_x}{\partial x}\,dz \qquad (12.4–13)$$

Write

$$(v_x)_0^\delta = u_s \qquad (12.4–14)$$

Combine equations (12.4–13) and (12.4–14) to give

$$(v_z v_x)_0^\delta = -u_s \int_0^\delta \frac{\partial v_x}{\partial x}\,dz \qquad (12.4–15)$$

Substitute equation (12.4–15) into equation (12.4–12) to give

$$\int_0^\delta \frac{\partial v_x^2}{\partial x}\,dz - u_s \int_0^\delta \frac{\partial v_x}{\partial x}\,dz = \int_0^\delta \eta \frac{\partial^2 v_x}{\partial z^2}\,dz \qquad (12.4–16)$$

$$\int_0^\delta \frac{\partial}{\partial x}(u_s - v_x)v_x\,dz = -\int_0^\delta \eta \frac{\partial^2 v_x}{\partial z^2}\,dz \qquad (12.4–17)$$

or

$$\int_0^\delta \frac{\partial}{\partial x}(u_s - v_x)v_x\,dz = \left(-\eta \frac{\partial v_x}{\partial z}\right)_0^\delta \qquad (12.4–18)$$

For Newtonian fluids the shear stress R_{zx} is related to velocity gradient dv_x/dz by equation (1.3–7)

$$R_{zx} = -\eta \rho \frac{dv_x}{dz} \qquad (1.3–7)$$

The shear stress at the wall of the plate R_w can be written as

$$R_w = -\eta\rho\frac{\partial v_x}{\partial z}\bigg|_{z=0} \qquad (12.4\text{–}19)$$

The shear stress R_{zx} at the boundary layer $z = \delta$ is zero.
Therefore

$$\frac{R_w}{\rho} = \left(-\eta\frac{\partial v_x}{\partial z}\right)_0^{\delta} \qquad (12.4\text{–}20)$$

Combine equations (12.4–18) and (12.4–20) to give

$$\int_0^{\delta} \frac{\partial}{\partial x}(u_s - v_x)v_x\,\mathrm{d}z = \frac{R_w}{\rho} \qquad (12.4\text{–}21)$$

Equation (12.4–21) is the von Karman integral equation for the boundary layer flow of a Newtonian liquid over a horizontal flat plate with zero pressure gradient. It is true for both laminar and turbulent flow boundary layers. The von Karman integral equation can be used to calculate the thickness of the boundary layer for both laminar and turbulent flow.

A Reynolds number for boundary layer flow can be defined as

$$N_{RE} = \frac{\rho u_s x}{\mu} \qquad (12.4\text{–}22)$$

where x is the distance away from the initial point of impact of the liquid with the flat plate as shown in Figure (12.4–1). At $N_{RE} = 3 \times 10^5$ to 3×10^6 the flow starts to become turbulent. However, even after full turbulence develops, a laminar sublayer remains in the immediate region of the plate surface.

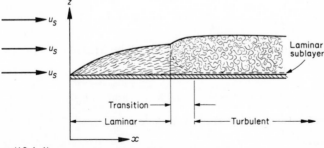

Figure (12.4–1)
Boundary layer over a horizontal flat plate.

12.5   Boundary layer thickness of a Newtonian liquid in steady laminar flow over a horizontal flat plate

Assume that the velocity profile in the laminar flow boundary layer of a Newtonian liquid on a flat plate obeys the equation

$$v_x = Az + Bz^2 + Cz^3 \tag{12.5-1}$$

where A, B and C are constants.

Equation (12.5-1) gives the point linear velocity of the liquid v_x at any vertical distance z from the plate.

Solve equation (12.5-1) for the following boundary layer conditions:

B.C.1 at $z = 0$, $v_x = 0$ and R_w = constant

B.C.2 at $z = \delta$, $v_x = u_s$ and $R_{zx} = 0$

The shear stress R_w for laminar flow over a flat plate is given by equation (12.4-20).

$$\frac{R_w}{\rho} = \left(-\eta \frac{\partial v_x}{\partial z} \right)^{\delta}_0 \tag{12.4-20}$$

Since R_w = constant at $z = 0$

$$\frac{\partial^2 v_x}{\partial z^2} = 0 \quad \text{at} \quad z = 0.$$

The shear stress R_{zx} is given by equation (1.3-7)

$$R_{zx} = -\eta \rho \frac{dv_x}{dz} \tag{1.3-7}$$

Since $R_{zx} = 0$ at $z = \delta$

$$\frac{\partial v_x}{\partial z} = 0 \quad \text{at} \quad z = \delta$$

Thus the boundary layer conditions can be written as follows:

B.C.1 at $z = 0$, $v_x = 0$ and $\partial^2 v_x/\partial z^2 = 0$

B.C.2 at $z = \delta$, $v_x = u_s$ and $\partial v_x/\partial z = 0$

Differentiate equation (12.5-1) to give

$$\frac{\partial v_x}{\partial z} = A + 2Bz + 3Cz^2 \tag{12.5-2}$$

and

$$\frac{\partial^2 v_x}{\partial z^2} = 2B + 6Cz \qquad (12.5\text{--}3)$$

From the boundary condition B.C.1, the constant $B = 0$. From the boundary condition B.C.2 the constants

$$A = \frac{3u_s}{2\delta} \qquad (12.5\text{--}4)$$

and

$$C = \frac{-u_s}{2\delta^3} \qquad (12.5\text{--}5)$$

Combine equations (12.5–1), (12.5–4) and (12.5–5) to give

$$\frac{v_x}{u_s} = \frac{3z}{2\delta} - \frac{1}{2}\left(\frac{z}{\delta}\right)^3 \qquad (12.5\text{--}6)$$

Differentiate equation (12.5–6) with respect to z at constant δ, i.e. at a particular distance x along the flat plate to give

$$\frac{\partial v_x}{\partial z} = u_s\left(\frac{3}{2\delta} - \frac{3z^2}{2\delta^3}\right) \qquad (12.5\text{--}7)$$

Therefore

$$\left.\frac{\partial v_x}{\partial z}\right|_{z=0} = \frac{3u_s}{2\delta} \qquad (12.5\text{--}8)$$

Differentiate equation (12.5–6) with respect to x at constant z to give

$$\frac{\partial v_x}{\partial x} = u_s\left[-\frac{3z}{2\delta^2}\left(\frac{d\delta}{dx}\right) + \frac{3z^3}{2\delta^4}\left(\frac{d\delta}{dx}\right)\right] \qquad (12.5\text{--}9)$$

Integrate equation (12.5–9) with respect to z over the boundary layer to give

$$\int_0^\delta \frac{\partial v_x}{\partial x}\,dz = u_s\left[-\frac{3z^2}{4\delta^2}\left(\frac{d\delta}{dx}\right) + \frac{3z^4}{8\delta^4}\left(\frac{d\delta}{dx}\right)\right]_0^\delta \qquad (12.5\text{--}10)$$

which becomes

$$\int_0^\delta \frac{\partial v_x}{\partial x}\,dz = -\frac{3u_s}{8}\left(\frac{d\delta}{dx}\right) \qquad (12.5\text{--}11)$$

Square equation (12.5–6) to give

$$v_x^2 = u_s^2\left(\frac{9z^2}{4\delta^2} + \frac{z^6}{4\delta^6} - \frac{3z^4}{2\delta^4}\right) \qquad (12.5\text{–}12)$$

Differentiate equation (12.5–12) with respect to x at a constant z to give

$$\frac{\partial v_x^2}{\partial x} = u_s^2\left[-\frac{9z^2}{2\delta^3}\left(\frac{d\delta}{dx}\right) - \frac{3z^6}{2\delta^7}\left(\frac{d\delta}{dx}\right) + \frac{6z^4}{\delta^5}\left(\frac{d\delta}{dx}\right) \right] \qquad (12.5\text{–}13)$$

Integrate equation (12.5–13) with respect to z over the boundary layer thickness δ to give

$$\int_0^\delta \frac{\partial v_x^2}{\partial x}\, dz = -\frac{18u_s^2}{35}\left(\frac{d\delta}{dx}\right) \qquad (12.5\text{–}14)$$

Equation (12.4–21) is the von Karman integral equation for the boundary layer flow of a Newtonian liquid over a horizontal flat plate with zero pressure gradient.

$$\int_0^\delta \frac{\partial}{\partial x}(u_s - v_x)v_x\, dz = \frac{R_w}{\rho} \qquad (12.4\text{–}21)$$

Substitute equations (12.5–11) and (12.5–13) into equation (12.4–21) to give

$$\frac{39u_s^2}{280}\left(\frac{d\delta}{dx}\right) = \frac{R_w}{\rho} \qquad (12.5\text{–}15)$$

Combine equations (12.4–20), (12.5–8) and (12.5–15) to give

$$\frac{39u_s^2}{280}\left(\frac{d\delta}{dx}\right) = \frac{3\eta u_s}{2\delta} \qquad (12.5\text{–}16)$$

which can be rewritten as

$$\delta\, d\delta = \frac{140\eta}{13u_s}\, dx \qquad (12.5\text{–}17)$$

Integrate equation (12.5–17) to

$$\frac{\delta^2}{2} = \frac{140\eta x}{13u_s} + C \qquad (12.5\text{–}18)$$

Since the vertical height or thickness of the boundary layer $\delta = 0$ at $x = 0$, the constant $C = 0$.

Therefore

$$\delta = 4.64 \sqrt{\frac{\eta x}{u_s}} \qquad (12.5\text{--}19)$$

Equation (12.5–19) gives the boundary layer thickness δ at any distance x along a horizontal flat plate for the laminar flow boundary layer.

For the laminar flow boundary layer on a horizontal flat plate

$$\delta \propto x^{0.5} \qquad (12.5\text{--}20)$$

Example (12.5–1)

A Newtonian liquid of density 1000 kg/m^3 and dynamic viscosity 0.01 kg/(s m) flows steadily over a horizontal flat plate with a stream velocity of 1.0 m/s. Calculate the Reynolds number and the boundary layer thickness in m at a distance 0.1 m along the plate from the initial point of impact of the liquid with the plate.

Calculations:

$$\text{Reynolds number } N_{RE} = \frac{\rho u_s x}{\mu} \qquad (12.4\text{--}22)$$

$$\rho = 1000 \text{ kg/m}^3$$

$$u_s = 1.0 \text{ m/s}$$

$$x = 0.1 \text{ m}$$

$$\mu = 0.01 \text{ kg/(s m)}$$

$$N_{RE} = \frac{(1000 \text{ kg/m}^3)(1.0 \text{ m/s})(0.1 \text{ m})}{0.01 \text{ kg/(s m)}}$$

$$= 1 \times 10^4$$

thus the liquid is in laminar flow at this point
boundary layer thickness for laminar flow

$$\delta = 4.64 \sqrt{\frac{\eta x}{u_s}} \qquad (12.5\text{--}19)$$

$$\eta = \frac{\mu}{\rho} = \frac{0.01 \text{ kg/(s m)}}{1000 \text{ kg/m}^3}$$

$$= 1 \times 10^{-5} \text{ m}^2/\text{s}$$

$$x = 0.1 \text{ m}$$

$$u_s = 1.0\,\text{m/s}$$

$$\frac{\eta x}{u_s} = \frac{(1 \times 10^{-5}\,\text{m}^2/\text{s})(0.1\,\text{m})}{1.0\,\text{m/s}}$$

$$= 1 \times 10^{-6}\,\text{m}^2$$

$$\sqrt{\frac{\eta x}{u_s}} = 1 \times 10^{-3}\,\text{m}$$

$$\delta = (4.64)(1 \times 10^{-3}\,\text{m})$$

$$= 0.00464\,\text{m}$$

12.6   Boundary layer thickness of a Newtonian liquid in steady turbulent flow over a horizontal flat plate

In this treatment it is assumed that the stream velocity u_s of the Newtonian liquid is sufficiently high so that the laminar flow region is small enough to be neglected. It is assumed that for all practical purposes, the boundary layer can be regarded as already turbulent at $x = 0$, the initial point of impact of the liquid with the flat plate.

The velocity profile equation for steady turbulent flow of a Newtonian fluid through a pipe of circular cross-section can be written either as

$$\frac{v_x}{v_{max}} = \left(\frac{2z}{d_i}\right)^{1/7} \tag{2.8-2}$$

or as

$$\frac{v_x}{v_{max}} = \left(\frac{z}{r}\right)^{1/7} \tag{12.6-1}$$

where z is the distance from the pipe wall and r and d_i are the radius and diameter of the pipe respectively.

Assume that the velocity profile in the turbulent flow boundary layer of a Newtonian liquid on a flat plate obeys the equation

$$\frac{v_x}{u_s} = \left(\frac{z}{\delta}\right)^{1/7} \tag{12.6-2}$$

where δ is the vertical height or thickness of the boundary layer at any distance x along the horizontal flat plate.

Solve equation (12.6–2) for the following boundary layer conditions:

$$\text{B.C.1 at } z = 0, \qquad v_x = 0$$

$$\text{B.C.2 at } z = \delta, \qquad v_x = u_s$$

Differentiate equation (12.6–2) with respect to x at constant z to give

$$\frac{\partial v_x}{\partial x} = \frac{-u_s z^{1/7} \, \delta^{-8/7}}{7} \left(\frac{d\delta}{dx}\right) \tag{12.6–3}$$

Integrate equation (12.6–3) with respect to z over the boundary layer to give

$$\int_0^\delta \frac{\partial v_x}{\partial x} \, dz = \left(\frac{-u_s z^{8/7} \, \delta^{-8/7}}{8}\right)_0^\delta \left(\frac{d\delta}{dx}\right) \tag{12.6–4}$$

which becomes

$$\int_0^\delta \frac{\partial v_x}{\partial x} \, dz = \frac{-u_s}{8} \left(\frac{d\delta}{dx}\right) \tag{12.6–5}$$

Square equation (12.6–2) to give

$$v_x^2 = u_s^2 \left(\frac{z}{\delta}\right)^{2/7} \tag{12.6–6}$$

Differentiate equation (12.6–6) with respect to x at a constant z to give

$$\frac{\partial v_x^2}{\partial x} = \frac{-2u_s^2 z^{2/7} \, \delta^{-9/7}}{7} \left(\frac{d\delta}{dx}\right) \tag{12.6–7}$$

Integrate equation (12.6–7) with respect to z over the boundary layer thickness δ to give

$$\int_0^\delta \frac{\partial v_x^2}{\partial x} \, dz = \frac{-2u_s^2}{9} \left(\frac{d\delta}{dx}\right) \tag{12.6–8}$$

Equation (12.4–21) is the von Karman integral equation for the boundary layer flow of a Newtonian liquid over a horizontal flat plate with zero pressure gradient.

$$\int_0^\delta \frac{\partial}{\partial x}(u_s - v_x)v_x \, dz = \frac{R_w}{\rho} \tag{12.4–21}$$

Substitute equations (12.6–4) and (12.6–8) into equation (12.4–21) to give

$$\frac{7u_s^2}{72}\left(\frac{d\delta}{dx}\right) = \frac{R_w}{\rho} \qquad (12.6–9)$$

The shear stress R_w for turbulent flow over a flat plate is given by the empirical Blasius equation which can be written as

$$\frac{R_w}{\rho u_s^2} = \frac{0.0228}{(\rho u_s \delta/\mu)^{0.25}} \qquad (12.6–10)$$

Combine equations (12.6–9) and (12.6–10) to give

$$\delta^{0.25}\, d\delta = 0.235\left(\frac{\mu}{\rho u_s}\right)^{0.25} dx \qquad (12.6–11)$$

Integrate equation (12.6–11) to

$$\delta = \frac{0.376x}{(\rho u_s x/\mu)^{0.2}} + C \qquad (12.6–12)$$

Since the vertical height or thickness of the boundary layer $\delta = 0$ at $x = 0$, the constant $C = 0$.

Therefore

$$\delta = \frac{0.376x}{(\rho u_s x/\mu)^{0.2}} \qquad (12.6–13)$$

Since the Reynolds number for boundary layer flow is defined as

$$N_{RE} = \frac{\rho u_s x}{\mu} \qquad (12.4–22)$$

equation (12.6–13) can be written as

$$\delta = \frac{0.376x}{N_{RE}^{0.2}} \qquad (12.6–14)$$

Equation (12.6–14) gives the boundary layer thickness δ at any distance x along a horizontal flat plate for the turbulent flow boundary layer.

For the turbulent flow boundary layer on a horizontal flat plate

$$\delta \propto x^{0.8} \qquad (12.6–15)$$

For the laminar flow boundary layer on a horizontal flat plate

$$\delta \propto x^{0.5} \qquad (12.6–16)$$

Thus the rate of increase of thickness of the boundary layer is greater in the turbulent than in the laminar region.

12.7   Steady vertical laminar flow of a Newtonian liquid

Let x and z be the horizontal and vertical downward directions respectively. Let there be no pressure gradient in the z directions and no fluid flow in the x and y directions.

The following relationships hold:

for gravitational acceleration

$$g_x = 0, \qquad g_y = 0, \qquad g_z = g$$

for velocity

$$v_x = 0, \qquad v_y = 0, \qquad v_z = v_z(x)$$

for steady flow

$$\frac{\partial v_z}{\partial t} = 0, \qquad \frac{\partial v_z}{\partial z} = 0, \qquad \frac{\partial v_z}{\partial y} = 0$$

for pressure gradient

$$\frac{\partial P}{\partial z} = 0$$

For these conditions the expanded modified Navier Stokes equations, equations (12.1–6), (12.1–7) and (12.1–8) reduce to

$$0 = \mu \frac{\partial^2 v_z}{\partial x^2} + \rho g \qquad (12.7\text{–}1)$$

Integrate equation (12.7–1) to

$$\frac{dv_z}{dx} = -\frac{\rho g x}{\mu} + C_1 \qquad (12.7\text{–}2)$$

and then to

$$v_z = -\frac{\rho g x^2}{2\mu} + C_1 x + C_2 \qquad (12.7\text{–}3)$$

where C_1 and C_2 are constants.

Equation (12.7–3) gives the point linear velocity v_z in the vertical z direction in terms of position x in the horizontal x direction for a Newtonian liquid of constant density and dynamic viscosity.

12.8   Steady laminar flow of a Newtonian liquid film down a vertical wall

Let the liquid film have a thickness b. Equation (12.7–3) gives the point linear velocity v_z in the vertical z direction in terms of position x in the horizontal x direction for a Newtonian liquid of constant density ρ and dynamic viscosity μ.

$$v_z = -\frac{\rho g x^2}{2\mu} + C_1 x + C_2 \qquad (12.7\text{–}3)$$

The velocity gradient at any point in the liquid film is given by equation (12.7–2).

$$\frac{dv_z}{dx} = -\frac{\rho g x}{\mu} + C_1 \qquad (12.7\text{–}2)$$

Solve equations (12.7–2) and (12.7–3) for the following boundary conditions:

$$\text{B.C.1 at } x = 0, \qquad v_z = 0$$

$$\text{B.C.2 at } x = b, \qquad R_{xz} = 0$$

The shear stress R_{xz} in the Newtonian liquid film is given by equation (1.3–7) written in the form

$$R_{xz} = -\eta\rho\frac{dv_z}{dx} \qquad (12.8\text{–}1)$$

Since the shear stress $R_{xz} = 0$ at the liquid gas interface $x = b$, the boundary conditions can be written as follows:

$$\text{B.C.1 at } x = 0, \qquad v_z = 0$$

$$\text{B.C.2 at } x = b, \qquad dv_z/dx = 0$$

From the boundary condition B.C.1, the constant $C_2 = 0$. From the boundary condition B.C.2 the constant

$$C_1 = \frac{\rho g b}{\mu} \qquad (12.8\text{–}2)$$

Substitute equation (12.8–2) into equation (12.7–3) to give

$$v_z = \frac{\rho g b^2}{\mu}\left(\frac{x}{b} - \frac{x^2}{2b^2}\right) \qquad (12.8\text{–}3)$$

Equation (12.8–3) gives the point linear velocity v_z in the vertical z direction at any horizontal position x in the vertical film of Newtonian liquid of constant density and dynamic viscosity falling under the influence of gravity.

The mean linear velocity u in the vertical z direction is given by the equation

$$u = \frac{1}{b} \int_0^b v_z \, dx \qquad (12.8\text{–}4)$$

Substitute equation (12.8–3) into equation (12.8–4) and carry out the integration to give

$$u = \frac{\rho g b^2}{3\mu} \qquad (12.8\text{–}5)$$

The volumetric flow rate of a film of Newtonian liquid in steady laminar flow down a vertical wall under the influence of gravity is

$$Q = \frac{\rho g b^3 \Delta y}{3\mu} \qquad (12.8\text{–}6)$$

where Δy is the width of the film in the horizontal y direction and b is the thickness of the film in the horizontal x direction.

REFERENCES

(1) Bennett, C. O., and Myers, J. E., *Momentum, Heat and Mass Transfer*, p. 91, New York, McGraw-Hill Book Co. Inc., 1962.

(2) Ibid., p. 113.

(3) Bird, R. B., Stewart, W. E., and Lightfoot, E. N., *Transport Phenomena*, p. 80, New York, John Wiley and Sons, Inc., 1960.

(4) Ibid., p. 142.

(5) Rohsenow, W. M., and Choi, H. Y., *Heat, Mass and Momentum Transfer*, p. 48, Englewood Cliffs, New Jersey, Prentice-Hall Inc., 1961.

(6) Welty, J. R., Wilson, R. E., and Wicks, C. E., *Fundamentals of Momentum, Heat and Mass Transfer*, p. 120, New York, John Wiley and Sons, Inc., 1969.

13
Applications of modified Navier Stokes equations in horizontal cylindrical coordinates

13.1 The modified Navier Stokes equations in horizontal cylindrical coordinates

The vector form of the modified Navier Stokes equations for momentum transfer is

$$\rho \frac{D\mathbf{v}}{Dt} = \mu \nabla^2 \mathbf{v} - \nabla P + \rho \mathbf{g} \qquad (12.1\text{--}5)$$

Equation (12.1–5) holds for Newtonian liquids. The modified Navier Stokes equations can be written in horizontal cylindrical coordinates as follows:

$$\rho \left(\frac{\partial v_r}{\partial t} + v_r \frac{\partial v_r}{\partial r} + \frac{v_\theta}{r} \frac{\partial v_r}{\partial \theta} - \frac{v_\theta^2}{r} + v_x \frac{\partial v_r}{\partial x} \right)$$

$$= -\frac{\partial P}{\partial r} + \mu \left\{ \frac{\partial}{\partial r} \left[\frac{1}{r} \frac{\partial}{\partial r}(rv_r) \right] + \frac{1}{r^2} \frac{\partial^2 v_r}{\partial \theta^2} - \frac{2}{r^2} \frac{\partial v_\theta}{\partial \theta} + \frac{\partial^2 v_r}{\partial x^2} \right\} + \rho g_r$$

$$(13.1\text{--}1)$$

$$\rho \left(\frac{\partial v_\theta}{\partial t} + v_r \frac{\partial v_\theta}{\partial r} + \frac{v_\theta}{r} \frac{\partial v_\theta}{\partial \theta} + \frac{v_r v_\theta}{r} + v_x \frac{\partial v_\theta}{\partial x} \right)$$

$$= -\frac{1}{r} \frac{\partial P}{\partial \theta} + \mu \left\{ \frac{\partial}{\partial r} \left[\frac{1}{r} \frac{\partial}{\partial r}(rv_\theta) \right] \right. \qquad (13.1\text{--}2)$$

$$\left. + \frac{1}{r^2} \frac{\partial^2 v_\theta}{\partial \theta^2} + \frac{2}{r^2} \frac{\partial v_r}{\partial \theta} + \frac{\partial^2 v_\theta}{\partial x^2} \right\} + \rho g_\theta$$

$$\rho\left(\frac{\partial v_x}{\partial t} + v_r\frac{\partial v_x}{\partial r} + \frac{v_\theta}{r}\frac{\partial v_x}{\partial \theta} + v_x\frac{\partial v_x}{\partial x}\right)$$

$$= -\frac{\partial P}{\partial x} + \mu\left[\frac{1}{r}\frac{\partial}{\partial r}\left(r\frac{\partial v_x}{\partial r}\right) + \frac{1}{r^2}\frac{\partial^2 v_x}{\partial \theta^2} + \frac{\partial^2 v_x}{\partial x^2}\right] + \rho g_x \quad (13.1\text{--}3)$$

where r is the radial distance from the horizontal axis in the x direction and θ is the angle the radius r makes with a fixed line which is perpendicular to the x axis.

13.2 Steady horizontal flow of a Newtonian liquid with no angular or radial velocity

For this system the following relationships hold:
for gravitational acceleration

$$g_x = 0, \qquad g_\theta = 0$$

for velocity

$$v_r = 0, \qquad v_\theta = 0, \qquad v_x = v_x(r)$$

for steady flow

$$\frac{\partial v_x}{\partial t} = 0, \qquad \frac{\partial v_x}{\partial x} = 0$$

for pressure gradient

$$\frac{\partial P}{\partial \theta} = 0$$

For these conditions the modified Navier Stokes equations in horizontal cylindrical coordinates, equations (13.1–1), (13.1–2) and (13.1–3) reduce to

$$0 = -\frac{\partial P}{\partial r} + \rho g_r \quad (13.2\text{--}1)$$

and

$$0 = \frac{\mu}{r}\frac{\partial}{\partial r}\left(r\frac{\partial v_x}{\partial r}\right) - \frac{\partial P}{\partial x} \quad (13.2\text{--}2)$$

Integrate equation (13.2–2) to

$$r\frac{dv_x}{dr} = \frac{1}{\mu}\left(\frac{dP}{dx}\right)\frac{r^2}{2} + C_1 \quad (13.2\text{--}3)$$

and then to

$$v_x = \frac{1}{\mu}\left(\frac{dP}{dx}\right)\frac{r^2}{4} + C_1 \ln r + C_2 \qquad (13.2\text{--}4)$$

where C_1 and C_2 are constants.

Equation (13.2–4) gives the point linear velocity v_x in the horizontal x direction in terms of position r in the radial direction for a Newtonian liquid of constant density and dynamic viscosity in steady laminar flow.

13.3 Steady laminar flow of a Newtonian liquid in a horizontal pipe

Equation (13.2–4) gives the point linear velocity v_x in the horizontal x direction in terms of position r in the radial direction for a Newtonian liquid in steady laminar flow.

$$v_x = \frac{1}{\mu}\left(\frac{dP}{dx}\right)\frac{r^2}{4} + C_1 \ln r + C_2 \qquad (13.2\text{--}4)$$

Equation (13.2–3) gives the velocity gradient dv_x/dr in terms of position r in the radial direction

$$r\frac{dv_x}{dr} = \frac{1}{\mu}\left(\frac{dP}{dx}\right)\frac{r^2}{2} + C_1 \qquad (13.2\text{--}3)$$

Solve equations (13.2–3) and (13.2–4) for the following boundary conditions:

$$\text{B.C.1 at } r = 0, \qquad dv_x/dr = 0$$

$$\text{B.C.2 at } r = d_i/2, \qquad v_x = 0$$

to give

$$C_1 = 0 \qquad (13.3\text{--}1)$$

and

$$C_2 = -\frac{1}{\mu}\left(\frac{dP}{dx}\right)\frac{d_i^2}{16} \qquad (13.3\text{--}2)$$

Substitute equations (13.3–1) and (13.3–2) into equation (13.2–4) to give

$$v_x = \frac{1}{\mu}\left(\frac{dP}{dx}\right)\left(\frac{r^2}{4} - \frac{d_i^2}{16}\right) \qquad (13.3\text{--}3)$$

Let the pressure drop be ΔP over a pipe length L so that the pressure gradient is given by the expression

$$-\frac{dP}{dx} = \frac{\Delta P}{L} \qquad (13.3\text{--}4)$$

Since the pressure P decreases with distance x the term $-dP/dx$ is positive.

Combine equations (13.3–3) and (13.3–4) to give

$$v_x = \left(\frac{\Delta P}{L}\right)\left(\frac{d_i^2}{16\mu}\right)\left[1 - \left(\frac{2r}{d_i}\right)^2\right] \qquad (2.7\text{--}3)$$

Equation (2.7–3) was previously derived by considering the shear stress R_{rx} at a radial distance r for steady laminar flow in a pipe.

Equation (2.7–3) gives the point linear velocity v_x at any radial distance r in a horizontal pipe of circular cross-section for a Newtonian liquid of constant density and dynamic viscosity in steady laminar flow.

13.4 Steady laminar flow of a Newtonian liquid in a horizontal concentric annulus

Equation (13.2–4) gives the point linear velocity v_x in the horizontal x direction in terms of position r in the radial direction for a Newtonian liquid in steady laminar flow.

$$v_x = \frac{1}{\mu}\left(\frac{dP}{dx}\right)\frac{r^2}{4} + C_1 \ln r + C_2 \qquad (13.2\text{--}4)$$

Consider a horizontal annulus of length L, inner diameter γd_i, and outer diameter d_i. Let the pressure drop be ΔP so that the pressure gradient is given by the expression

$$-\frac{dP}{dx} = \frac{\Delta P}{L} \qquad (13.3\text{--}4)$$

Solve equation (13.2–4) for the following boundary conditions:

$$\text{B.C.1 at } r = d_i/2, \qquad v_x = 0$$
$$\text{B.C.2 at } r = \gamma d_i/2, \qquad v_x = 0$$

to give

$$C_1 = -\left(\frac{dP}{dx}\right)\left(\frac{d_i^2}{16\mu}\right)\frac{(1-\gamma^2)}{\ln(1/\gamma)} \qquad (13.4\text{--}1)$$

and

$$C_2 = -\left(\frac{dP}{dx}\right)\left(\frac{d_i^2}{16\mu}\right)\left[1 - (1 - \gamma^2)\frac{\ln(d_i/2)}{\ln(1/\gamma)}\right] \qquad (13.4\text{--}2)$$

Substitute equations (13.3–4), (13.4–1) and (13.4–2) into equation (13.2–4) to give

$$v_x = \left(\frac{\Delta P}{L}\right)\left(\frac{d_i^2}{16\mu}\right)\left[1 - \left(\frac{2r}{d_i}\right)^2 + (1 - \gamma^2)\frac{\ln(2r/d_i)}{\ln(1/\gamma)}\right] \qquad (13.4\text{--}3)$$

Equation (13.4–3) gives the point linear velocity v_x at any radial distance r in a horizontal concentric annulus for a Newtonian liquid of constant density and dynamic viscosity in steady laminar flow.

When $\gamma = 0$, equation (13.4–3) becomes

$$v_x = \left(\frac{\Delta P}{L}\right)\left(\frac{d_i^2}{16\mu}\right)\left[1 - \left(\frac{2r}{d_i}\right)^2\right] \qquad (2.7\text{--}3)$$

which is the velocity profile equation for a Newtonian liquid in laminar flow in a horizontal pipe of circular cross-section.

The mean linear velocity u in the annulus is given by the equation

$$u = \frac{2\pi \int_{\gamma d_i/2}^{d_i/2} v_x r\,dr}{2\pi \int_{\gamma d_i/2}^{d_i/2} r\,dr} \qquad (13.4\text{--}4)$$

Substitute equation (13.4–3) into equation (13.4–4) and carry out the integration to give

$$u = \left(\frac{\Delta P}{L}\right)\left(\frac{d_i^2}{32\mu}\right)\left[\frac{(1 - \gamma^4)}{(1 - \gamma^2)} - \frac{(1 - \gamma^2)}{\ln(1/\gamma)}\right] \qquad (13.4\text{--}5)$$

Equation (13.4–5) is obtained using the standard integral

$$\int r\ln\left(\frac{2r}{d_i}\right) = \frac{r^2}{4}\left[2\ln\left(\frac{2r}{d_i}\right) - 1\right] \qquad (13.4\text{--}6)$$

The volumetric flow rate through the annulus

$$Q = \frac{\pi d_i^2}{4}(1 - \gamma^2)u \qquad (13.4\text{--}7)$$

or in terms of equation (13.4–5)

$$Q = \left(\frac{\Delta P}{L}\right)\left(\frac{\pi d_i^4}{128\mu}\right)\left[(1 - \gamma^4) - \frac{(1 - \gamma^2)^2}{\ln(1/\gamma)}\right] \qquad (13.4\text{--}8)$$

Example (13.4–1)

Calculate the steady volumetric flow rate for the following system:
(1) A horizontal concentric annulus with an inner diameter of 0.07 m and an outer diameter of 0.1 m.
(2) A pressure gradient in the annulus of 4000 $(N/m^2)/m$.
(3) A liquid dynamic viscosity of 1.0 N s/m^2 and a liquid density of 1100 kg/m^3.

Calculations:
  volumetric flow rate

$$Q = \left(\frac{\Delta P}{L}\right)\left(\frac{\pi d_i^4}{128\mu}\right)\left[(1 - \gamma^4) - \frac{(1 - \gamma^2)^2}{\ln(1/\gamma)}\right] \qquad (13.4\text{--}8)$$

$$d_i = 0.1 \text{ m}$$

$$\mu = 1.0 \text{ N s/m}^2$$

$$\Delta P/L = 4000 \text{ (N/m}^2)/\text{m}$$

$$\left(\frac{\Delta P}{L}\right)\left(\frac{\pi d_i^4}{128\mu}\right) = \frac{[4000 \text{ (N/m}^2)/\text{m}](3.142)(0.1 \text{ m})^4}{(128)(1.0 \text{ N s/m}^2)}$$

$$= 9.819 \times 10^{-3} \text{ m}^3/\text{s}$$

$$\gamma = 0.7$$

$$\gamma^2 = 0.49$$

$$\gamma^4 = 0.2401$$

$$(1 - \gamma^4) = 0.7599$$

$$(1 - \gamma^2)^2 = 0.2601$$

$$1/\gamma = \frac{1}{0.7} = 1.429$$

$$\ln 1.429 = 0.3569$$

$$\frac{(1 - \gamma^2)^2}{\ln(1/\gamma)} = \frac{0.2601}{0.3569} = 0.7288$$

$$Q = (9.819 \times 10^{-3} \text{ m}^3/\text{s})(0.7599 - 0.7288)$$

$$= 3.054 \times 10^{-4} \text{ m}^3/\text{s}$$

13.5   Steady laminar flow of a Newtonian liquid in a horizontal concentric annulus with the inner cylinder moving at a constant velocity with no applied pressure gradient[1]

Equation (13.2–4) gives the point linear velocity v_x in the horizontal x direction in terms of position r in the radial direction for a Newtonian liquid in steady laminar flow.

$$v_x = \frac{1}{\mu}\left(\frac{dP}{dx}\right)\frac{r^2}{4} + C_1 \ln r + C_2 \qquad (13.2\text{–}4)$$

For zero pressure gradient, equation (13.2–4) becomes

$$v_x = C_1 \ln r + C_2 \qquad (13.5\text{–}1)$$

Consider a horizontal annulus of inner diameter γd_i and outer diameter d_i. Let the inner wall move with a steady linear velocity v_o.

Solve equation (13.5–1) for the following boundary conditions:

$$\text{B.C.1 at } r = d_i/2, \qquad v_x = 0$$

$$\text{B.C.2 at } r = \gamma d_i/2, \qquad v_x = v_o$$

to give

$$C_1 = \frac{v_o}{\ln \gamma} \qquad (13.5\text{–}2)$$

and

$$C_2 = -\frac{v_o \ln (d_i/2)}{\ln \gamma} \qquad (13.5\text{–}3)$$

Substitute equations (13.5–2) and (13.5–3) into equation (13.5–1) to give

$$v_x = v_o \frac{\ln (2r/d_i)}{\ln \gamma} \qquad (13.5\text{–}4)$$

Equation (13.5–4) gives the point linear velocity v_x at any radial distance r in a horizontal concentric annulus for a Newtonian liquid of constant density and dynamic viscosity in steady laminar flow with the inner cylinder moving at a constant velocity with no applied pressure gradient.

The mean linear velocity u in the annulus is given by the equation

$$u = \frac{2\pi \int_{\gamma d_i/2}^{d_i/2} v_x r \, dr}{2\pi \int_{\gamma d_i/2}^{d_i/2} r \, dr} \qquad (13.4\text{–}4)$$

Substitute equation (13.5–4) into equation (13.4–4) and carry out the integration to give

$$u = \frac{v_o}{2}\left[\frac{1}{\ln{(1/\gamma)}} - \frac{2\gamma^2}{(1 - \gamma^2)}\right] \qquad (13.5\text{–}5)$$

Equation (13.5–5) is obtained using the standard integral

$$\int r \ln\left(\frac{2r}{d_i}\right) = \frac{r^2}{4}\left[2\ln\left(\frac{2r}{d_i}\right) - 1\right] \qquad (13.4\text{–}6)$$

The volumetric flow rate through the annulus

$$Q = \frac{\pi d_i^2}{4}(1 - \gamma^2)u \qquad (13.4\text{–}7)$$

or in terms of equation (13.5–5)

$$Q = \left(\frac{\pi d_i^2 v_o}{8}\right)\left[\frac{(1 - \gamma^2)}{\ln{(1/\gamma)}} - 2\gamma^2\right] \qquad (13.5\text{–}6)$$

13.6 Steady laminar flow of a Newtonian liquid in a horizontal annulus with the inner cylinder moving at a constant velocity with an applied pressure gradient

Equation (13.2–4) gives the point linear velocity v_x in the horizontal x direction in terms of position r in the radial direction for a Newtonian liquid in steady laminar flow.

$$v_x = \frac{1}{\mu}\left(\frac{dP}{dx}\right)\frac{r^2}{4} + C_1 \ln r + C_2 \qquad (13.2\text{–}4)$$

Consider a horizontal annulus of length L, inner diameter γd_i, and outer diameter d_i. Let the inner wall move with a steady linear velocity v_o. Also let the pressure drop be ΔP so that the pressure gradient is given by the expression

$$-\frac{dP}{dx} = \frac{\Delta P}{L} \qquad (13.3\text{–}4)$$

Solve equation (13.2–4) for the following boundary conditions:

B.C.1 at $r = d_i/2$, $v_x = 0$

B.C.2 at $r = \gamma d_i/2$, $v_x = v_o$

to give

$$C_1 = -\left(\frac{dP}{dx}\right)\left(\frac{d_i^2}{16\mu}\right)\frac{(1-\gamma^2)}{\ln(1/\gamma)} + \frac{v_o}{\ln\gamma} \qquad (13.6-1)$$

and

$$C_2 = -\left(\frac{dP}{dx}\right)\left(\frac{d_i^2}{16\mu}\right)\left[1 - (1-\gamma^2)\frac{\ln(d_i/2)}{\ln(1/\gamma)}\right] - \frac{v_o\ln(d_i/2)}{\ln\gamma} \qquad (13.6-2)$$

Substitute equations (13.3–4), (13.6–1) and (13.6–2) into equation (13.2–4) to give

$$v_x = \left(\frac{\Delta P}{L}\right)\left(\frac{d_i^2}{16\mu}\right)\left[1 - \left(\frac{2r}{d_i}\right)^2 + (1-\gamma^2)\frac{\ln(2r/d_i)}{\ln(1/\gamma)}\right] - v_o\frac{\ln(2r/d_i)}{\ln(1/\gamma)}$$

$$(13.6-3)$$

Equation (13.6–3) gives the point linear velocity v_x at any radial distance r in a horizontal concentric annulus for a Newtonian liquid of constant density and dynamic viscosity in steady laminar flow with the inner cylinder moving at a constant velocity with an applied pressure gradient.

The mean linear velocity u in the annulus is given by the equation

$$u = \frac{2\pi\int_{\gamma d_i/2}^{d_i/2} v_x r\, dr}{2\pi\int_{\gamma d_i/2}^{d_i/2} r\, dr} \qquad (13.4-4)$$

Substitute equation (13.6–3) into equation (13.4–4) and carry out the integration to give

$$u = \left(\frac{\Delta P}{L}\right)\left(\frac{d_i^2}{32\mu}\right)\left[\frac{(1-\gamma^4)}{(1-\gamma^2)} - \frac{(1-\gamma^2)}{\ln(1/\gamma)}\right] + \frac{v_o}{2}\left[\frac{1}{\ln(1/\gamma)} - \frac{2\gamma^2}{(1-\gamma^2)}\right]$$

$$(13.6-4)$$

The volumetric flow rate through the annulus

$$Q = \frac{\pi d_i^2}{4}(1-\gamma^2)u \qquad (13.4-7)$$

or in terms of equation (13.6–4)

$$Q = \left(\frac{\Delta P}{L}\right)\left(\frac{\pi d_i^4}{128\mu}\right)\left[(1-\gamma^4) - \frac{(1-\gamma^2)^2}{\ln(1/\gamma)}\right]$$

$$+ \left(\frac{\pi d_i^2 v_o}{8}\right)\left[\frac{(1-\gamma^2)}{\ln(1/\gamma)} - 2\gamma^2\right] \qquad (13.6-5)$$

It is interesting to note that the volumetric flow rate Q given by equation (13.6–5) is simply the sum of Q given by equation (13.4–8) for flow under pressure in an ordinary horizontal concentric annulus and Q given by equation (13.5–6) for flow in an annulus with the inner cylinder moving at a constant velocity with no applied pressure gradient.

13.7 Unsteady laminar flow of a Newtonian liquid in a horizontal pipe[3]

For this system the following relationships hold:
for gravitational acceleration

$$g_x = 0, \qquad g_\theta = 0$$

for velocity

$$v_r = 0, \qquad v_\theta = 0, \qquad v_x = v_x(r, t)$$

for pressure gradient

$$\frac{\partial P}{\partial \theta} = 0$$

For these conditions the modified Navier Stokes equations in horizontal cylindrical coordinates, equations (13.1–1), (13.1–2) and (13.1–3) reduce to

$$\rho \frac{\partial v_x}{\partial t} = \frac{\mu}{r} \frac{\partial}{\partial r}\left(r \frac{\partial v_x}{\partial r}\right) - \frac{\partial P}{\partial x} \qquad (13.7\text{–}1)$$

which can also be written as

$$\frac{\partial v_x}{\partial t} = \frac{\eta}{r} \frac{\partial}{\partial r}\left(r \frac{\partial v_x}{\partial r}\right) - \frac{1}{\rho} \frac{\partial P}{\partial x} \qquad (13.7\text{–}2)$$

in terms of the kinematic viscosity $\eta = \mu/\rho$.

Consider a long horizontal pipe of length L and inside diameter d_i filled with a Newtonian liquid of constant density ρ and constant dynamic viscosity μ at rest. At a time $t = 0$, apply a pressure drop ΔP across the length L of the pipe sufficient to cause the liquid to flow in laminar motion. The point velocity $v_x = v_x(r, t)$ at any position r in the radial direction and any time t can be obtained by integrating equation (13.7–2).

The pressure gradient is given by the expression

$$-\frac{\mathrm{d}P}{\mathrm{d}x} = \frac{\Delta P}{L} \tag{13.3-4}$$

Rewrite equation (13.7–2) in terms of equation (13.3–4) as

$$\frac{\partial v_x}{\partial t} = \frac{\eta}{r}\frac{\partial}{\partial r}\left(r\frac{\partial v_x}{\partial r}\right) + \frac{\Delta P}{\rho L} \tag{13.7-3}$$

Write the velocity v_x as

$$v_x(r, t) = \bar{v}_x(r) - v_t(r, t) \tag{13.7-4}$$

where $\bar{v}_x(r)$ is the steady point velocity attained at any position r in the radial direction after a time $t \to \infty$ and $v_t(r, t)$ is a transient velocity function which varies with the time t; $v_t(r, t) \to 0$ as $t \to \infty$.

Since $\mathrm{d}v_x/\mathrm{d}t \to 0$ and $v_x(r, t) \to \bar{v}_x(r)$ as $t \to 0$, equation (13.7–3) can be written for steady flow as

$$0 = \frac{\eta}{r}\frac{\partial}{\partial r}\left(r\frac{\partial \bar{v}_x}{\partial r}\right) + \frac{\Delta P}{\rho L} \tag{13.7-5}$$

Equation (13.7–5) can be solved for the following boundary conditions:

$$\text{B.C.1 at } r = 0, \qquad \mathrm{d}v_x/\mathrm{d}_r = 0$$
$$\text{B.C.2 at } r = d_i/2, \qquad v_x = 0$$

to give

$$\bar{v}_x = \left(\frac{\Delta P}{L}\right)\left(\frac{d_i^2}{16\mu}\right)\left[1 - \left(\frac{2r}{d_i}\right)^2\right] \tag{13.7-6}$$

Equation (13.7–6) is equivalent to equation (2.7–3) which has already been derived. Equation (2.7–3) gives the velocity profile for a Newtonian liquid in steady laminar flow in a horizontal pipe of circular cross-section.

Differentiate equation (13.7–4) with respect to t, to give

$$\frac{\partial v_x}{\partial t} = -\frac{\partial v_t}{\partial t} \tag{13.7-7}$$

and with respect to r, to give

$$\frac{\partial v_x}{\partial r} = \frac{\partial \bar{v}_x}{\partial r} - \frac{\partial v_t}{\partial r} \tag{13.7-8}$$

Combine equations (13.7–3) and (13.7–8) to give

$$\frac{\partial v_x}{\partial t} = \eta\left[\frac{1}{r}\frac{\partial}{\partial r}\left(r\frac{\partial \bar{v}_x}{\partial r} - r\frac{\partial v_t}{\partial r}\right)\right] + \frac{\Delta P}{\rho L} \qquad (13.7–9)$$

Integrate equation (13.7–5) to give

$$\frac{\partial \bar{v}_x}{\partial r} = -\left(\frac{\Delta P}{\rho L\eta}\right)\frac{r}{2} \qquad (13.7–10)$$

which can be combined with equation (13.7–9) to give

$$\frac{\partial v_x}{\partial t} = \eta\left[-\frac{1}{r}\frac{\partial}{\partial r}\left(\frac{\Delta P r^2}{2\rho L\eta}\right)\right] - \frac{\eta}{r}\frac{\partial}{\partial r}\left(r\frac{\partial v_t}{\partial r}\right) + \frac{\Delta P}{\rho L} \quad (13.7–11)$$

which simplifies to

$$\frac{\partial v_x}{\partial t} = -\frac{\eta}{r}\frac{\partial}{\partial r}\left(r\frac{\partial v_t}{\partial r}\right) \qquad (13.7–12)$$

Combine equations (13.7–7) and (13.7–12) and write

$$\frac{\partial v_t}{\partial t} = \frac{\eta}{r}\frac{\partial}{\partial r}\left(r\frac{\partial v_t}{\partial r}\right) \qquad (13.7–13)$$

which can be written in the expanded form as

$$\frac{\partial v_t}{\partial t} = \eta\left(\frac{\partial^2 v_t}{\partial r^2} + \frac{1}{r}\frac{\partial v_t}{\partial r}\right) \qquad (13.7–14)$$

Equation (13.7–14) is a partial differential equation which can be solved by the separation of variables technique as follows. Assume a solution

$$v_t(r, t) = Z(r)T(t) \qquad (13.7–15)$$

Differentiate equation (13.7–15) with respect to t, to give

$$\frac{\partial v_t}{\partial t} = Z\frac{dT}{dt} \qquad (13.7–16)$$

and with respect to r, to give

$$\frac{\partial v_t}{\partial r} = T\frac{dZ}{dr} \qquad (13.7–17)$$

and

$$\frac{\partial^2 v_t}{\partial r^2} = T\frac{d^2 Z}{dr^2} \qquad (13.7–18)$$

Substitute equations (13.7–16), (13.7–17) and (13.7–18) into equation (13.7–14) to give

$$Z\frac{dT}{dt} = \eta\left(T\frac{d^2Z}{dr^2} + \frac{T}{r}\frac{dZ}{dr}\right) \qquad (13.7\text{–}19)$$

which can also be written as

$$\frac{1}{T}\frac{dT}{dt} = \eta\left[\frac{1}{Z}\frac{d^2Z}{dr^2} + \frac{1}{Z}\left(\frac{1}{r}\frac{dZ}{dr}\right)\right] \qquad (13.7\text{–}20)$$

The left hand side of equation (13.7–20) is independent of r and the right hand side is independent of t. This is only possible if both sides are independent of r and t. Thus each side is equal to a constant which can be written as $-\alpha$. Combine this constant respectively with the left and right hand sides of equation (13.7–20) to give the following two ordinary differential equations:

$$\frac{dT}{dt} + \alpha T = 0 \qquad (13.7\text{–}21)$$

and

$$\frac{d^2Z}{dr^2} + \frac{1}{r}\frac{dZ}{dr} + \left(\frac{\alpha}{\eta}\right)Z = 0 \qquad (13.7\text{–}22)$$

Write $\alpha/\eta = \beta^2$ and substitute into equation (13.7–22) to give

$$\frac{d^2Z}{dr^2} + \frac{1}{r}\frac{dZ}{dr} + \beta^2 Z = 0 \qquad (13.7\text{–}23)$$

which can also be written as

$$r^2\frac{d^2Z}{dr^2} + r\frac{dZ}{dr} + \beta^2 r^2 Z = 0 \qquad (13.7\text{–}24)$$

Write[6] $s = \beta r$ and substitute into equation (13.7–24) to give

$$s^2\frac{d^2Z}{ds^2} + s\frac{dZ}{ds} + s^2 Z = 0 \qquad (13.7\text{–}25)$$

which is the well known Bessel equation of zero order.

The solution to equation (13.7–21) is

$$T = C_1 e^{-\alpha t} \qquad (13.7\text{–}26)$$

where C_1 is a constant.

Equation (13.7–24) can be solved using the method of Frobenius[4] to give

$$Z = C_2 J_o(s) + C_3 Y_o(s) \tag{13.7–27}$$

where C_2 and C_3 are constants.

$J_o(s)$ is the most commonly encountered Bessel function.[7] It represents the converging power series expansion

$$J_o(s) = \sum_{n=0}^{\infty} \frac{(-1)^n}{(n!)^2} \left(\frac{s}{2}\right)^{2n} = 1 - \frac{s^2}{2^2} + \frac{s^4}{2^2 \cdot 4^2} - \frac{s^6}{2^2 \cdot 4^2 \cdot 6^2} + \cdots$$

$$\tag{13.7–28}$$

$Y_o(s)$ is also a converging function. $J_o(s)$ and $Y_o(s)$ are graphically illustrated in Figures (13.7–1) and (13.7–2) respectively. Thus the

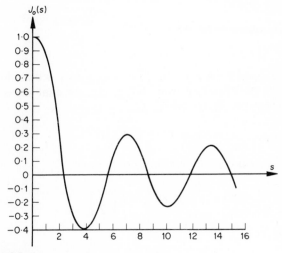

Figure (13.7–1)
Graphical illustration of Bessel function $J_o(s)$.

solution of equation (13.7–14) can be written in the form of equation (13.7–15) as

$$v_t = C_1 e^{-\alpha t}[C_2 J_o(s) + C_3 Y_o(s)] \tag{13.7–29}$$

Solve equation (13.7–29) for the following boundary conditions:

B.C.1 at $r = 0$, v_t is finite

B.C.2 at $r = d_i/2$, $v_t = 0$

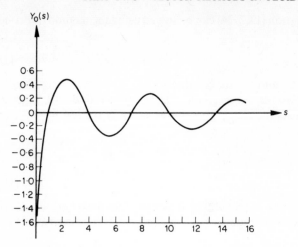

Figure (13.7–2)
Graphical illustration of Bessel function $Y_o(s)$.

Since $s = \beta r$, $s = 0$ when $r = 0$. $Y_o(s) \to -\infty$ as $s \to 0$ as shown in Figure (13.7–2). Therefore from B.C.1, the constant $C_3 = 0$ and equation (13.7–29) can be written as

$$v_t = C_1 C_2 \, e^{-\alpha t} J_o(s) \qquad (13.7\text{–}30)$$

For B.C.2, equation (13.7–30) can be written as

$$0 = C_1 C_2 \, e^{-\alpha t} J_o\left(\frac{\beta d_i}{2}\right) \qquad (13.7\text{–}31)$$

which reduces to

$$J_o\left(\frac{\beta d_i}{2}\right) = 0 \qquad (13.7\text{–}32)$$

Thus $\beta d_i/2$ has the values of s where $J_o(s)$ cuts the abscissa in Figure (13.7–1).
   Write

$$\frac{\beta d_i}{2} = \lambda_n \qquad (13.7\text{–}33)$$

where $\lambda_1 = 2.405$, $\lambda_2 = 5.520$, $\lambda_3 = 8.654$, $\lambda_4 = 11.792$, $\lambda_5 = 14.931$, $\lambda_6 = 18.071$ and so on.

Rewrite equation (13.7–33) as

$$\beta = \frac{2\lambda_n}{d_i} \tag{13.7–34}$$

Since $\alpha = \eta\beta^2$, substitute this into equation (13.7–34) to give

$$\alpha = \eta\left(\frac{2\lambda_n}{d_i}\right)^2 \tag{13.7–35}$$

Since $s = \beta r$, equation (13.7–34) can be written as

$$s = \frac{2\lambda_n r}{d_i} \tag{13.7–36}$$

Substitute equations (13.7–35) and (13.7–36) into equation (13.7–30) to give

$$v_t = \sum_{n=1}^{n=\infty} C_{1n}C_{2n} \exp\left[-\eta\left(\frac{2\lambda_n}{d_i}\right)^2 t\right] J_o\left(\frac{2\lambda_n r}{d_i}\right) \tag{13.7–37}$$

where all the values of n are included.

Simplify equation (13.7–37) by writing $C = C_1 C_2$ to give

$$v_t = \sum_{n=1}^{n=\infty} C_n \exp\left[-\eta\left(\frac{2\lambda_n}{d_i}\right)^2 t\right] J_o\left(\frac{2\lambda_n r}{d_i}\right) \tag{13.7–38}$$

where the values of C_n are arbitrary constants.

At time $t = 0$, $v_x = 0$ and from equation (13.7–4) $v_t = \bar{v}_x$. Thus at $t = 0$, equation (13.7–38) can be written as

$$\bar{v}_x = \sum_{n=1}^{n=\infty} C_n J_o\left(\frac{2\lambda_n r}{d_i}\right) \tag{13.7–39}$$

Evaluate the constants C_n using the orthogonality properties of Bessel functions[5,8] as follows. Multiply both sides of equation (13.7–39) by $rJ_o(2\lambda_m r/d_i)$ and integrate from $r = 0$ to $r = d_i/2$ with respect to r to give

$$\int_0^{d_i/2} r\bar{v}_x J_o\left(\frac{2\lambda_m r}{d_i}\right) dr = \sum_{n=1}^{n=\infty} C_n \int_0^{d_i/2} r J_o\left(\frac{2\lambda_n r}{d_i}\right) J_o\left(\frac{2\lambda_m r}{d_i}\right) dr \tag{13.7–40}$$

and then

$$\int_0^{d_i/2} r\bar{v}_x J_o\left(\frac{2\lambda_m r}{d_i}\right) dr = C_n \frac{d_i^2}{8} J_1^2(\lambda_n) \tag{13.7–41}$$

The derivation of equation (13.7–41) is based on the following two properties:

$$\int_0^{d_i/2} J_o\left(\frac{2\lambda_n r}{d_i}\right) J_o\left(\frac{2\lambda_m r}{d_i}\right) dr = 0 \qquad \text{for } n \neq m$$

and

$$\int_0^{d_i/2} r J_o\left(\frac{2\lambda_n r}{d_i}\right) J_o\left(\frac{2\lambda_m r}{d_i}\right) dr = \frac{d_i^2}{8}[J_1(\lambda_n)]^2 \delta_{nm}$$

where δ_{nm} is the Kronecker delta[8]; $\delta_{nm} = +1$ if $n = m$, and $\delta_{nm} = 0$ if $n \neq m$.

Rewrite equation (13.7–41) as

$$C_n = \frac{8}{d_i^2 J_1^2(\lambda_n)} \int_0^{d_i/2} r \bar{v}_x J_o\left(\frac{2\lambda_n r}{d_i}\right) dr \qquad (13.7\text{–}42)$$

where $\bar{v}_x$ is given by equation (13.7–6).

The integral on the right hand side of equation (13.7–42) can be evaluated[2] to give

$$C_n = \left(\frac{\Delta P}{L}\right)\left(\frac{d_i^2}{16\mu}\right)\left[\frac{8}{\lambda_n^3 J_1(\lambda_n)}\right] \qquad (13.7\text{–}43)$$

Combine equations (13.7–38) and (13.7–43) to give

$$v_t = 8\left(\frac{\Delta P}{L}\right)\left(\frac{d_i^2}{16\mu}\right) \sum_{n=1}^{n=\infty} \frac{\exp\left[-\eta(2\lambda_n/d_i)^2 t\right] J_o(2\lambda_n r/d_i)}{\lambda_n^3 J_1(\lambda_n)} \qquad (13.7\text{–}44)$$

Equation (13.7–44) is the solution to the transient part of equation (13.7–4) and equation (13.7–6) is the solution to the steady part. Combine equations (13.7–4), (13.7–6) and (13.7–44) to give the full solution

$$v_x = \left(\frac{\Delta P}{L}\right)\left(\frac{d_i^2}{16\mu}\right)\left\{1 - \left(\frac{2r}{d_i}\right)^2 - 8\sum_{n=1}^{n=\infty} \frac{\exp\left[-\eta(2\lambda_n/d_i)^2 t\right] J_o(2\lambda_n r/d_i)}{\lambda_n^3 J_1(\lambda_n)}\right\}$$

$$(13.7\text{–}45)$$

Equation (13.7–45) gives the point linear velocity v_x at a time t at any radial distance r in a horizontal pipe of circular cross-section for a Newtonian liquid of constant density and dynamic viscosity in unsteady laminar flow after being initially at rest.

As the time $t \to \infty$, equation (13.7–45) reduces to

$$v_x = \left(\frac{\Delta P}{L}\right)\left(\frac{d_i^2}{16\mu}\right)\left[1 - \left(\frac{2r}{d_i}\right)^2\right]$$ (2.7–3)

Equation (2.7–3) gives the point linear velocity v_x at any radial distance r in a horizontal pipe of circular cross-section for a Newtonian liquid of constant density and dynamic viscosity in steady laminar flow. Unsteady state velocity profiles can be calculated from equation (13.7–45) for various times t. Some of these are plotted[9] in Figure (13.7–3) in relation to the steady state velocity

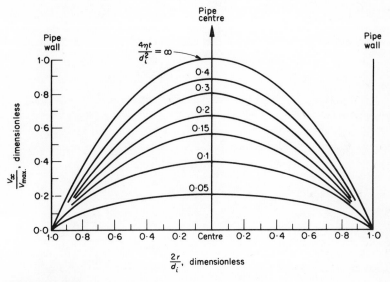

Figure (13.7–3)
Velocity profiles for a Newtonian liquid in a pipe as a function of the time from rest.

profile for a Newtonian liquid of constant density and dynamic viscosity in laminar flow in a horizontal pipe of circular cross-section.

REFERENCES

(1) Bird, R. B., Stewart, W. E., and Lightfoot, E. N., *Transport Phenomena*, p. 65, New York, John Wiley and Sons, Inc., 1960.
(2) Ibid., p. 129.

(3) Holland, F. A., and Chapman, F. S., *Pumping of Liquids*, p. 45, New York, Reinhold Publishing Corporation, 1966.
(4) Hildebrand, F. B., *Advanced Calculus for Applications*, p. 129, Englewood Cliffs, New Jersey, Prentice-Hall, Inc., 1962.
(5) Ibid., p. 448.
(6) Miller, K. S., *Partial Differential Equations in Engineering Problems*, p. 179, Englewood Cliffs, New Jersey, Prentice-Hall, Inc., 1953.
(7) Ibid., p. 205.
(8) Ibid., p. 206.
(9) Szymanski, P., J. Math. Pures Appl., **11**, 67 (1932).

14
Applications of modified Navier Stokes equations in vertical cylindrical coordinates

14.1 The modified Navier Stokes equations in vertical cylindrical coordinates

The vector form of the modified Navier Stokes equations for momentum transfer is

$$\rho\frac{D\mathbf{v}}{Dt} = \mu\nabla^2\mathbf{v} - \nabla P + \rho\mathbf{g} \qquad (12.1-5)$$

Equation (12.1–5) holds for Newtonian liquids. The modified Navier Stokes equations can be written in vertical cylindrical coordinates as follows:

$$\rho\left(\frac{\partial v_r}{\partial t} + v_r\frac{\partial v_r}{\partial r} + \frac{v_\theta}{r}\frac{\partial v_r}{\partial \theta} - \frac{v_\theta^2}{r} + v_z\frac{\partial v_r}{\partial z}\right)$$

$$= -\frac{\partial P}{\partial r} + \mu\left\{\frac{\partial}{\partial r}\left[\frac{1}{r}\frac{\partial}{\partial r}(rv_r)\right] + \frac{1}{r^2}\frac{\partial^2 v_r}{\partial \theta^2} - \frac{2}{r^2}\frac{\partial v_\theta}{\partial \theta} + \frac{\partial^2 v_r}{\partial z^2}\right\} + \rho g_r$$

$$(14.1-1)$$

$$\rho\left(\frac{\partial v_\theta}{\partial t} + v_r\frac{\partial v_\theta}{\partial r} + \frac{v_\theta}{r}\frac{\partial v_\theta}{\partial \theta} + \frac{v_r v_\theta}{r} + v_z\frac{\partial v_\theta}{\partial z}\right)$$

$$= -\frac{1}{r}\frac{\partial P}{\partial \theta} + \mu\left\{\frac{\partial}{\partial r}\left[\frac{1}{r}\frac{\partial}{\partial r}(rv_\theta)\right]\right. \qquad (14.1-2)$$

$$\left. + \frac{1}{r^2}\frac{\partial^2 v_\theta}{\partial \theta^2} + \frac{2}{r^2}\frac{\partial v_r}{\partial \theta} + \frac{\partial^2 v_\theta}{\partial z^2}\right\} + \rho g_\theta$$

$$\rho\left(\frac{\partial v_z}{\partial t} + v_r\frac{\partial v_z}{\partial r} + \frac{v_\theta}{r}\frac{\partial v_z}{\partial \theta} + v_z\frac{\partial v_z}{\partial z}\right)$$

$$= -\frac{\partial P}{\partial z} + \mu\left[\frac{1}{r}\frac{\partial}{\partial r}\left(r\frac{\partial v_z}{\partial r}\right) + \frac{1}{r^2}\frac{\partial^2 v_z}{\partial \theta^2} + \frac{\partial^2 v_z}{\partial z^2}\right] + \rho g_z \qquad (14.1\text{--}3)$$

where r is the radial distance from the vertical axis in the z direction and θ is the angle the radius r makes with a fixed line which is perpendicular to the z axis.

14.2   Steady vertical flow of a Newtonian liquid with no angular or radial velocity

For this system where z is taken as the vertical downward direction, the following relationships hold:

for gravitational acceleration

$$g_r = 0, \qquad g_\theta = 0, \qquad g_z = g$$

for velocity

$$v_r = 0, \qquad v_\theta = 0, \qquad v_z = v_z(r), \qquad \frac{\partial v_z}{\partial \theta} = 0$$

for steady flow

$$\frac{\partial v_z}{\partial t} = 0, \qquad \frac{\partial v_z}{\partial z} = 0$$

for pressure gradient

$$\frac{\partial P}{\partial r} = 0, \qquad \frac{\partial P}{\partial \theta} = 0$$

For these conditions the modified Navier Stokes equations in vertical cylindrical coordinates, equations (14.1–1), (14.1–2) and (14.1–3) reduce to

$$0 = -\frac{\partial P}{\partial z} + \frac{\mu}{r}\frac{\partial}{\partial r}\left(r\frac{\partial v_z}{\partial r}\right) + \rho g \qquad (14.2\text{--}1)$$

The pressure gradient $-\partial P/\partial z$ would normally be very small. If this can be neglected equation (14.2–1) becomes

$$0 = \frac{\mu}{r}\frac{\partial}{\partial r}\left(r\frac{\partial v_z}{\partial r}\right) + \rho g \qquad (14.2\text{--}2)$$

Integrate equation (14.2–2) to

$$r\frac{dv_z}{dr} = -\left(\frac{\rho g}{\mu}\right)\frac{r^2}{2} + C_1 \qquad (14.2\text{–}3)$$

and then to

$$v_z = -\left(\frac{\rho g}{\mu}\right)\frac{r^2}{4} + C_1 \ln r + C_2 \qquad (14.2\text{–}4)$$

where C_1 and C_2 are constants.

Equation (14.2–4) gives the point linear velocity v_z in the vertical z direction in terms of position r in the radial direction for a Newtonian liquid of constant density and dynamic viscosity in steady laminar flow.

14.3 Steady laminar flow of a Newtonian liquid film down the outside of a vertical tube

Equation (14.2–4) gives the point linear velocity v_z in the vertical z direction in terms of position r in the radial r direction for a Newtonian liquid in steady laminar flow.

$$v_z = -\left(\frac{\rho g}{\mu}\right)\frac{r^2}{4} + C_1 \ln r + C_2 \qquad (14.2\text{–}4)$$

Equation (14.2–3) gives the velocity gradient dv_z/dr in terms of position r in the radial direction

$$r\frac{dv_z}{dr} = -\left(\frac{\rho g}{\mu}\right)\frac{r^2}{2} + C_1 \qquad (14.2\text{–}3)$$

Solve equations (14.2–3) and (14.2–4) for the following boundary conditions:

$$\text{B.C.1 at } r = d_o/2, \qquad v_z = 0$$

$$\text{B.C.2 at } r = \gamma d_o/2, \qquad dv_z/dr = 0$$

to give

$$C_1 = \left(\frac{\rho g}{\mu}\right)\frac{\gamma^2 d_o^2}{8} \qquad (14.3\text{–}1)$$

and

$$C_2 = \left(\frac{\rho g}{\mu}\right)\left(\frac{d_o^2}{16}\right)[1 - 2\gamma^2 \ln (d_o/2)] \qquad (14.3\text{–}2)$$

where d_o is the outside diameter of the vertical tube and γd_o is the outside diameter of the liquid film.

Substitute equations (14.3–1) and (14.3–2) into equation (14.2–4) to give

$$v_z = \left(\frac{\rho g}{\mu}\right)\left(\frac{d_o^2}{16}\right)\left[1 - \left(\frac{2r}{d_o}\right)^2 + 2\gamma^2 \ln\left(\frac{2r}{d_o}\right)\right] \qquad (14.3\text{–}3)$$

Equation (14.3–3) gives the point linear velocity v_z at any radial distance r on the outside of a vertical tube for a film of Newtonian liquid of constant density and dynamic viscosity falling freely in steady laminar flow.

The volumetric flow rate of the film

$$Q = 2\pi \int_{d_o/2}^{\gamma d_o/2} r v_z \, dr \qquad (14.3\text{–}4)$$

Substitute equation (14.3–3) into equation (14.3–4) and carry out the integration to give

$$Q = \left(\frac{\rho g}{\mu}\right)\left(\frac{\pi d_o^4}{128}\right)(-1 + 4\gamma^2 - 3\gamma^4 + 4\gamma^4 \ln \gamma) \qquad (14.3\text{–}5)$$

Equation (14.3–5) is obtained using the standard integral

$$\int r \ln\left(\frac{2r}{d_o}\right) = \frac{r^2}{4}\left[2 \ln\left(\frac{2r}{d_o}\right) - 1\right] \qquad (14.3\text{–}6)$$

Equation (14.3–5) gives the volumetric flow rate of a film of thickness b of Newtonian liquid falling freely in steady laminar flow on the outside of a vertical tube where b is related to γ by the equations

$$\gamma = \frac{b + (d_o/2)}{d_o/2} \qquad (14.3\text{–}7)$$

and

$$\gamma = 1 + \frac{2b}{d_o} \qquad (14.3\text{–}8)$$

Example (14.3–1)
  Calculate the steady volumetric flow rate for the following system:
(1) A film of thickness 3×10^{-3} m on the outside of a vertical tube of diameter 0.15 m.

(2) A liquid dynamic viscosity of 1.0 N s/m^2 and a liquid density of 1120 kg/m^3.
(3) A gravitational acceleration of 9.81 m/s^2.

Calculations:
  volumetric flow rate

$$Q = \left(\frac{\rho g}{\mu}\right)\left(\frac{\pi d_o^4}{128}\right)(-1 + 4\gamma^2 - 3\gamma^4 + 4\gamma^4 \ln \gamma) \quad (14.3\text{–}5)$$

$$\rho = 1120 \text{ kg/m}^3$$

$$g = 9.81 \text{ m/s}^2$$

$$\mu = 1.0 \text{ N s/m}^2$$

$$\frac{\rho g}{\mu} = \frac{(1120 \text{ kg/m}^3)(9.81 \text{ m/s}^2)}{(1.0 \text{ N s/m}^2)} = 10987.2 \text{ s}^{-1}\text{ m}^{-1}$$

$$\pi = 3.142$$

$$d_o = 0.15 \text{ m}$$

$$\frac{\pi d_o^4}{128} = \frac{(3.142)(0.15 \text{ m})^4}{128} = 1.24269 \times 10^{-5} \text{ m}^4$$

$$\gamma = 1 + \frac{2b}{d_o}$$

$$b = 3 \times 10^{-3} \text{ m}, \qquad d_o = 0.15 \text{ m}$$

$$2b/d_o = 0.04$$

$$\gamma = 1.04, \qquad \gamma^2 = 1.08160, \qquad \gamma^4 = 1.16986$$

$$4\gamma^2 = 4.32640$$

$$3\gamma^4 = 3.50958$$

$$4\gamma^4 = 4.67944$$

$$\ln \gamma = 0.039220$$

$$4\gamma^4 \ln \gamma = 0.18353$$

$$(-1 + 4\gamma^2 - 3\gamma^4 + 4\gamma^4 \ln \gamma) = 3.5 \times 10^{-4}$$

$$Q = (10\,990 \text{ s}^{-1}\text{ m}^{-1})(1.243 \times 10^{-5} \text{ m}^4)(3.5 \times 10^{-4})$$

$$= 4.779 \times 10^{-5} \text{ m}^3/\text{s}$$

This answer is extremely sensitive to the number of figures carried due to the subtraction of nearly equal quantities. Whether the number of figures required is justified by the accuracy of the data must be decided in each particular case.

14.4   Steady laminar rotational flow of a Newtonian liquid about a vertical axis with no vertical or radial velocity

For this system where z is taken as the vertical downward direction, the following relationships hold:
  for gravitational acceleration

$$g_r = 0, \qquad g_\theta = 0, \qquad g_z = g$$

  for velocity

$$v_r = 0, \qquad v_\theta = v_\theta(r), \qquad v_z = 0$$

$$\frac{\partial v_\theta}{\partial \theta} = 0, \qquad \frac{\partial v_\theta}{\partial z} = 0$$

  for steady flow

$$\frac{\partial v_\theta}{\partial t} = 0$$

  for pressure gradient

$$\frac{\partial P}{\partial \theta} = 0$$

For these conditions the modified Navier Stokes equations in vertical cylindrical coordinates, equations (14.1–1), (14.1–2) and (14.1–3) reduce to

$$\rho \frac{v_\theta^2}{r} = \frac{\partial P}{\partial r} \tag{14.4–1}$$

$$0 = \frac{\partial}{\partial r}\left[\frac{1}{r}\frac{\partial}{\partial r}(rv_\theta)\right] \tag{14.4–2}$$

$$0 = -\frac{\partial P}{\partial z} + \rho g \tag{14.4–3}$$

Integrate equation (14.4–2) to

$$\frac{1}{r}\frac{d}{dr}(rv_\theta) = C_1 \tag{14.4–4}$$

and then to

$$v_\theta = \frac{C_1 r}{2} + \frac{C_2}{r} \tag{14.4-5}$$

where C_1 and C_2 are constants.

Equation (14.4–5) gives the point linear velocity v_θ in the angular θ direction in terms of position r in the radial direction for a Newtonian liquid of constant density and dynamic viscosity in steady laminar flow.

14.5 Steady laminar rotational flow of a Newtonian liquid between coaxial vertical cylinders rotating with different angular velocities

Equation (14.4–5) gives the point linear velocity v_θ in the angular θ direction in terms of position r in the radial r direction for a Newtonian liquid in steady laminar flow.

$$v_\theta = \frac{C_1 r}{2} + \frac{C_2}{r} \tag{14.4-5}$$

Consider a vertical annulus of inner diameter γd_i and outer diameter d_i. Let the inner and outer walls move with steady rotational velocities of ω_1 and ω_2 radians per second.

Solve equation (14.4–5) for the following boundary conditions:

B.C.1 at $r = d_i/2$, $v_\theta = \omega_2 d_i/2$

B.C.2 at $r = \gamma d_i/2$, $v_\theta = \omega_1 \gamma d_i/2$

to give

$$C_1 = \frac{2(\omega_2 - \gamma^2 \omega_1)}{(1 - \gamma^2)} \tag{14.5-1}$$

and

$$C_2 = -\frac{\gamma^2 d_i^2 (\omega_2 - \omega_1)}{4(1 - \gamma^2)} \tag{14.5-2}$$

Substitute equations (14.5–1) and (14.5–2) into equation (14.4–5) to give

$$v_\theta = \frac{(\omega_2 - \gamma^2 \omega_1)r}{(1 - \gamma^2)} - \frac{\gamma^2 d_i^2 (\omega_2 - \omega_1)}{4r(1 - \gamma^2)} \tag{14.5-3}$$

Equation (14.5–3) gives the point linear velocity v_θ in the angular θ direction in terms of position r in a vertical annulus for a Newtonian liquid of constant density and dynamic viscosity in steady laminar flow with the inner and outer walls rotating at different velocities.

It can be shown[1] that the shear stress $R_{r\theta}$ at a position r in the radial direction for a Newtonian liquid of constant density and dynamic viscosity in steady rotational laminar flow with no radial velocity is given by the equation

$$R_{r\theta} = -\mu r \frac{\partial}{\partial r}\left(\frac{v_\theta}{r}\right) \tag{14.5–4}$$

Differentiate equation (14.5–3) to give

$$\frac{d}{dr}\left(\frac{v_\theta}{r}\right) = \frac{\gamma^2 d_i^2(\omega_2 - \omega_1)}{2r^3(1 - \gamma^2)} \tag{14.5–5}$$

Combine equations (14.5–4) and (14.5–5) to give

$$R_{r\theta} = -\frac{\mu\gamma^2 d_i^2(\omega_2 - \omega_1)}{2r^2(1 - \gamma^2)} \tag{14.5–6}$$

Equation (14.5–6) illustrates the principle of viscometers which consist of two vertical coaxial cylinders. Fluid viscosities are obtained from torque and rotational velocity measurements.

In Couette–Hatschek viscometers, the inner cylinder is stationary, i.e. $\omega_1 = 0$ and the outer cylinder rotates. In this case equation (14.5–6) reduces to

$$R_{r\theta} = -\frac{\mu\gamma^2 d_i^2 \omega_2}{2r^2(1 - \gamma^2)} \tag{14.5–7}$$

The shear stress at the outer cylinder

$$R_{r\theta}|_{r = d_i/2} = -\frac{2\mu\gamma^2\omega_2}{(1 - \gamma^2)} \tag{14.5–8}$$

The torque required to turn the outer cylinder

$$\tau = \pi d_i L(-R_{r\theta}|_{r = d_i/2})\frac{d_i}{2} \tag{14.5–9}$$

where L is the vertical height of the cylinders.

Combine equations (14.5–8) and (14.5–9) to give

$$\tau = \pi\mu L d_i^2 \omega_2\left(\frac{\gamma^2}{1 - \gamma^2}\right) \tag{14.5–10}$$

Equation (14.5–10) gives the relationship between torque, rotational velocity and dynamic viscosity in a Couette–Hatschek type viscometer.

In Stormer viscometers, the outer cylinder is stationary, i.e. $\omega_2 = 0$ and the inner cylinder rotates. In this case equation (14.5–6) reduces to

$$R_{r\theta} = \frac{\mu\gamma^2 d_i^2 \omega_1}{2r^2(1 - \gamma^2)} \qquad (14.5–11)$$

The shear stress at the inner cylinder

$$R_{r\theta}|_{r=\gamma d_i/2} = \frac{2\mu d_i^2 \omega_1}{(1 - \gamma^2)} \qquad (14.5–12)$$

The torque required to turn the inner cylinder

$$\tau = \pi\gamma d_i L(+ R_{r\theta}|_{r=\gamma d_i/2})\frac{\gamma d_i}{2} \qquad (14.5–13)$$

Combine equations (14.5–12) and (14.5–13) to give

$$\tau = \pi\mu L d_i^2 \omega_1 \left(\frac{\gamma^2}{1 - \gamma^2}\right) \qquad (14.5–14)$$

Equation (14.5–14) gives the relationship between torque, rotational velocity, and dynamic viscosity in a Stormer type viscometer.

Rotational viscometers can be calibrated for non-Newtonian fluids.[3]

14.6 Steady laminar rotational flow of a Newtonian liquid producing a parabolic vortex[2,4]

Equation (14.5–3) gives the point linear velocity v_θ in a vertical annulus for a Newtonian liquid in steady laminar flow with the inner and outer walls rotating at ω_1 and ω_2 radians per second respectively.

$$v_\theta = \frac{(\omega_2 - \gamma^2\omega_1)r}{(1 - \gamma^2)} - \frac{\gamma^2 d_i^2(\omega_2 - \omega_1)}{4r(1 - \gamma^2)} \qquad (14.5–3)$$

For the case of no inner cylinder $\gamma = 0$. Put $\omega = \omega_2$ and rewrite equation (14.5–3) as

$$v_\theta = \omega r \qquad (14.6–1)$$

For steady laminar rotational flow of a Newtonian liquid about a vertical axis with no vertical or radial velocity, the modified Navier Stokes equations in vertical cylindrical coordinates, equations (14.1–1) and (14.1–3), reduce respectively to

$$\rho \frac{v_\theta^2}{r} = \frac{\partial P}{\partial r} \tag{14.4–1}$$

and

$$0 = -\frac{\partial P}{\partial z} + \rho g \tag{14.4–3}$$

Combine equations (14.4–1) and (14.6–1) to give

$$\frac{\partial P}{\partial r} = \rho \omega^2 r \tag{14.6–2}$$

Write the differential pressure function as

$$dP = \frac{\partial P}{\partial r} \, dr + \frac{\partial P}{\partial z} \, dz \tag{14.6–3}$$

Substitute equations (14.4–3) and (14.6–2) into equation (14.6–3) to give

$$dP = \rho \omega^2 r \, dr + \rho g \, dz \tag{14.6–4}$$

which can be integrated to give

$$P = \frac{\rho \omega^2 r^2}{2} + \rho g z + C \tag{14.6–5}$$

where C is a constant.

Let $z = 0$ at $r = d_i/2$ as shown in Figure (14.6–1). At this point on the liquid surface, the pressure is atmospheric, i.e. $P = P_A$. Therefore in equation (14.6–5) the constant C is given by the equation

$$C = P_A - \frac{\rho \omega^2}{2}\left(\frac{d_i}{2}\right)^2 \tag{14.6–6}$$

Substitute equation (14.6–6) into equation (14.6–5) to give

$$P - P_A = \frac{\rho \omega^2}{2}\left[r^2 - \left(\frac{d_i}{2}\right)^2\right] + \rho g z \tag{14.6–7}$$

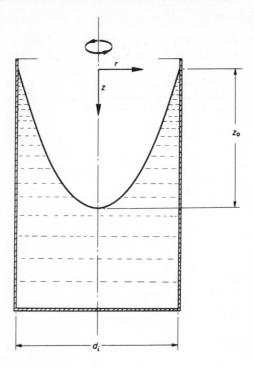

Figure (14.6–1)
Rotating Newtonian liquid producting a parabolic vortex.

Since the pressure P is atmospheric at every point on the free surface, equation (14.6–7) for the free surface can be written as

$$z = \frac{\omega^2}{2g}\left[\left(\frac{d_i}{2}\right)^2 - r^2\right]$$

(14.6–8)

which is the equation of a parabola. At $r = 0$

$$z_o = \frac{\omega^2 d_i^2}{8g}$$

(14.6–9)

Equation (14.6–8) gives the depth of the free surface at any radial position r for a Newtonian liquid of constant density and dynamic viscosity in steady laminar rotational flow.

REFERENCES

(1) Bird, R. B., Stewart, W. E., and Lightfoot, E. N., *Transport Phenomena*, p. 89, New York, John Wiley and Sons, Inc., 1960.
(2) Ibid., p. 96.
(3) Garcia-Borras, T., *Chem. Eng.*, **72**, No. 2 (1965).
(4) Kay, J. M., *An Introduction to Fluid Mechanics and Heat Transfer*, 2nd ed., p. 20, Cambridge, The University Press, 1963.

Conversion factors

area	$1\ \text{ft}^2$	$= 0.092903\ \text{m}^2$
density	$1\ \text{lb/ft}^3$	$= 16.018\ \text{kg/m}^3$
	$1\ \text{lb/UK gal}$	$= 99.779\ \text{kg/m}^3$
	$1\ \text{lb/US gal}$	$= 119.83\ \text{kg/m}^3$
dynamic viscosity	$1\ \text{cp}$	$= 0.001\ \text{N s/m}^2$
	$1\ \text{lb/(h ft)}$	$= 4.1338 \times 10^{-4}\ \text{N s/m}^2$
	$1\ \text{lb/(s ft)}$	$= 1.4882\ \text{N s/m}^2$
energy	$1\ \text{Btu}$	$= 1055.06\ \text{J}$
	$1\ \text{ft pdl}$	$= 0.042139\ \text{J}$
flow rate, mass per unit time	$1\ \text{lb/h}$	$= 1.2600 \times 10^{-4}\ \text{kg/s}$
flow rate, volume per unit time	$1\ \text{ft}^3/\text{s}$	$= 0.028317\ \text{m}^3/\text{s}$
	$1\ \text{ft}^3/\text{min}$	$= 4.7195 \times 10^{-4}\ \text{m}^3/\text{s}$
	$1\ \text{UK gal/min}$	$= 7.5766 \times 10^{-5}\ \text{m}^3/\text{s}$
	$1\ \text{US gal/min}$	$= 6.3089 \times 10^{-5}\ \text{m}^3/\text{s}$
heat capacity per unit mass	$1\ \text{Btu/(lb F)}$	$= 4186.8\ \text{J/(kg K)}$
kinematic viscosity	$1\ \text{ft}^2/\text{s}$	$= 0.092903\ \text{m}^2/\text{s}$
length	$1\ \text{ft}$	$= 0.3048\ \text{m}$
linear velocity	$1\ \text{ft/s}$	$= 0.3048\ \text{m/s}$
mass	$1\ \text{lb}$	$= 0.45359\ \text{kg}$
mass flow rate	$1\ \text{lb/(h ft}^2)$	$= 1.3562 \times 10^{-3}\ \text{kg/(s m}^2)$
pressure	$1\ \text{atm}$	$= 101\,325\ \text{N/m}^2$
	$1\ \text{pdl/ft}^2$	$= 1.4882\ \text{N/m}^2$
	$1\ \text{psi}$	$= 6894.8\ \text{N/m}^2$
pressure gradient	$1\ (\text{pdl/ft}^2)/\text{ft}$	$= 4.8824\ (\text{N/m}^2)/\text{m}$
power	$1\ \text{ft pdl/s}$	$= 0.04214\ \text{W}$
	$1\ \text{hp (British)}$	$= 745.7\ \text{W}$
	$1\ \text{ton refrigeration}$	$= 3516.9\ \text{W}$
specific volume	$1\ \text{ft}^3/\text{lb}$	$= 0.062428\ \text{m}^3/\text{kg}$
surface tension	$1\ \text{dyne/cm}$	$= 0.001\ \text{N/m}$
temperature difference	$1\ \text{F}$	$= 0.5556\ \text{K}$
volume	$1\ \text{ft}^3$	$= 0.028317\ \text{m}^3$
	$1\ \text{UK gal}$	$= 0.0045460\ \text{m}^3$
	$1\ \text{US gal}$	$= 0.0037853\ \text{m}^3$

APPENDIX

Further problems

(The numbers refer to relevant chapter and section)

(1.4–1) (a) Derive an expression for the apparent dynamic viscosity μ_a of a power law fluid.

(b) Derive an expression for the rate of change of apparent dynamic viscosity of a power law fluid with change in shear rate $\dot{\gamma}$.

(c) Calculate the apparent dynamic viscosity μ_a of a power law fluid if the consistency coefficient $K = 1.5\,(\text{N/m}^2)\,\text{s}^{0.5}$, the power law index $n = 0.5$, and the shear rate $\dot{\gamma} = 100\,\text{s}^{-1}$.

(d) Calculate the rate of change of μ_a with change in $\dot{\gamma}$ for the data in part (c).

(1.6–1) An incompressible fluid flows upwards in steady state in a cylindrical pipe at an angle θ with the horizontal. Assume that the head loss due to friction is negligible.

(a) Derive an expression for the pressure gradient in the pipe.

(b) Derive an expression for the length of pipe over which the pressure is reduced by half.

(c) Calculate the length of pipe L over which the pressure is reduced by half if the gravitational acceleration $g = 9.81\,\text{m/s}^2$, $\theta = 30°$, the liquid density $\rho = 1200\,\text{kg/m}^3$, and the initial pressure $P_1 = 200\,000\,\text{N/m}^2$.

(2.4–1) (a) Derive an expression for the mean linear velocity u of a liquid in steady state turbulent flow in a smooth cylindrical tube in terms of the pressure gradient $\Delta P/L$, the liquid density ρ, the inside diameter of the tube d_i and the liquid dynamic viscosity μ.

(b) Calculate u if

$$\Delta P/L = 528\,\text{N/m}^3$$

$$\rho = 1200\,\text{kg/m}^3$$

$$\mu = 0.01\,\text{N s/m}^2$$

and
$$d_i = 0.05\,\text{m}$$

(2.4–2) A liquid flows in steady state in a cylindrical pipe with a Reynolds number of 2000. Derive an expression for the liquid dynamic viscosity μ in terms of the pressure gradient $\Delta P/L$, the liquid density ρ, and the inside pipe diameter d_i.

(2.4–3) (a) Derive an expression for the pressure gradient $\Delta P/L$ for a liquid in steady state turbulent flow in a rough cylindrical pipe in terms of the liquid density ρ, the mean linear velocity u, the inside pipe diameter d_i and the roughness factor ε.
(b) Use this to calculate $\Delta P/L$ for

$$\rho = 1200 \text{ kg/m}^3$$
$$d_i = 0.0526 \text{ m}$$
$$u = 1.160 \text{ m/s}$$
$$\mu = 0.01 \text{ N s/m}^2$$

and
$$\varepsilon = 0.000045 \text{ m}$$

(2.4–4) A fluid of density ρ and dynamic viscosity μ flows in steady state in a cylindrical pipe of inside diameter d_i with a mean linear velocity u. Use dimensional analysis to derive an expression for the pressure gradient $\Delta P/L$ in terms of ρ, u, d_i and μ.

(2.5–1) Calculate the pressure gradient $\Delta P/L$ for a liquid in steady state turbulent flow in a coil of inside tube diameter $d_i = 0.02$ m and coil diameter $D_C = 2$ m if the liquid density $\rho = 1200 \text{ kg/m}^3$, the liquid dynamic viscosity $\mu = 0.001 \text{ N s/m}^2$ and the mean linear velocity $u = 2$ m/s.

(2.5–2) A liquid flows in steady state in a cylindrical pipe of inside diameter $d_i = 0.05$ m at a flow rate $Q = 2 \times 10^{-3} \text{ m}^3/\text{s}$. Calculate the head loss and the pressure drop for a sudden expansion to a pipe of inside diameter 0.1 m, if the liquid density $\rho = 1000 \text{ kg/m}^3$.

(2.7–1) The laminar flow velocity profile in a pipe for a Newtonian liquid in steady state flow is given by the equation

$$v_x = 2u\left[1 - \left(\frac{2r}{d_i}\right)^2\right]$$

where v_x is the point linear velocity at any radial position r. Use this to derive an expression for the velocity gradient at the pipe wall.

(2.7–2) A Newtonian liquid flows in steady state in a cylindrical pipe.
(a) Calculate point linear velocities v_x at the centre and at a radius of one quarter of the pipe diameter if the inside diameter of the pipe $d_i = 0.05$ m, the liquid dynamic viscosity $\mu = 0.15 \text{ N s/m}^2$, the liquid density $\rho = 1100 \text{ kg/m}^3$ and the pressure gradient $\Delta P/L = 3000 \text{ (N/m}^2)/\text{m}$.
(b) Calculate the volumetric flow rate Q through the pipe.

(2.7–3) Substitute equation (2.7–3) into equation (2.7–6) and integrate to give equation (2.7–7).

(2.7–4) Substitute equations (2.7–3) and (2.7–5) into equation (2.7–13) and integrate to give equation (2.7–14).

(2.8–1) Plot laminar and turbulent velocity profile curves respectively for steady state flow in a cylindrical pipe for a maximum point linear velocity $v_{max} = 5$ m/s using the radial positions $2r/d_i = 0$, 0.2, 0.4, 0.6 and 0.08.

(3.3–1) Calculate the pressure gradient $\Delta P/L$ for a time independent non-Newtonian fluid in steady state flow in a cylindrical tube if

the liquid density $\qquad \rho = 1000$ kg/m^3
the inside diameter of the tube $\qquad d_i = 0.08$ m
the mean linear velocity $\qquad u = 1$ m/s
the point pipe consistency coefficient $\qquad K'_p = 2\,(\text{N/m}^2)\,\text{s}^{0.5}$
and the flow behaviour index $\qquad n' = 0.5$

(3.6–1) Substitute the equation

$$R_{rx} = K\left(\frac{-\mathrm{d}v_x}{\mathrm{d}r}\right)^n$$

into equation

$$\frac{8u}{d_i} = \frac{32}{d_i^3}\int_0^{d_i/2} r^2\left(\frac{-\mathrm{d}v_x}{\mathrm{d}r}\right)\mathrm{d}r \qquad (2.10\text{–}4)$$

and integrate to show that the shear rate at a pipe wall for a power law fluid in steady state flow is

$$\dot{\gamma}_w = \left(\frac{8u}{d_i}\right)\left(\frac{3n+1}{4n}\right)$$

(3.6–2) Show that the mean shear rate $\dot{\gamma}_m$ for a power law fluid in steady state flow in a pipe is related to the shear rate at the wall $\dot{\gamma}_w$ by the equation

$$\dot{\gamma}_m = \dot{\gamma}_w\frac{n}{(n+1)}$$

where n is the power law index.

(3.6–3) For Newtonian fluids in which the dynamic viscosity μ is a function only of temperature, i.e. $\mu = f(T)$. The expression $N_{VIS} = \mu_w/\mu$ raised to some power is used to correct isothermal equations for non-isothermal conditions. Suggest an analogous correction for non-Newtonian power law fluids flowing in pipes in which the apparent dynamic viscosity μ_a is a function of shear stress R, shear rate $\dot{\gamma}$ and temperature T, i.e., $\mu_a = f(R, \dot{\gamma}, T)$.

(3.7–1) The laminar flow velocity profile in a pipe for a power law liquid in steady state flow is given by the equation

$$v_x = u\left(\frac{3n+1}{n+1}\right)\left[1 - \left(\frac{2r}{d_i}\right)^{(n+1)/n}\right]$$

where n is the power law index and u is the mean linear velocity. Use this to derive the expression

$$\left(\frac{-dv_x}{dr}\right) = \left(\frac{8u}{d_i}\right)\left(\frac{3n+1}{4n}\right)$$

for the velocity gradient at the pipe wall.

(4.1–1) Figure (1.6–2) diagrammatically represents the heads in a liquid flowing through a pipe. Redraw this diagram with a pump placed between points 1 and 2.

(4.2–1) Calculate the available net positive suction head $NPSH$ in a pumping system if the liquid density $\rho = 1200\,\text{kg/m}^3$, the liquid dynamic viscosity $\mu = 0.4\,\text{N s/m}^2$, the mean linear velocity $u = 1\,\text{m/s}$, the static head on the suction side $z_s = 3\,\text{m}$, the inside pipe diameter $d_i = 0.0526\,\text{m}$, the gravitational acceleration $g = 9.81\,\text{m/s}^2$, and the equivalent length on the suction side $\Sigma L_{es} = 5.0\,\text{m}$.
The liquid is at its boiling point. Neglect entrance and exit losses.

(4.3–1) A centrifugal pump is used to pump a liquid in steady turbulent flow through a smooth pipe from one tank to another. Develop an expression for the system total head Δh in terms of the static heads on the discharge and suction sides z_d and z_s respectively, the gas pressures above the tanks on the discharge and suction sides P_d and P_s respectively, the liquid density ρ, the liquid dynamic viscosity μ, the gravitational acceleration g, the total equivalent lengths on the discharge and suction sides ΣL_{ed} and ΣL_{es} respectively, and the volumetric flow rate Q.

(4.3–2) A system total head against mean linear velocity curve for a particular power law liquid in a particular pipe system can be represented by the equation

$$\Delta h = (0.03)(100^n)(u^n) + 4.0 \quad \text{for } u \leqslant 1.5\,\text{m/s}$$

where

Δh is the total head in m

u is the mean linear velocity in m/s

and

n is the power law index.

A centrifugal pump operates in this particular system with a total head against mean linear velocity curve represented by the equation

$$\Delta h = 8.0 - 0.2u - 1.0u^2 \quad \text{for } u \leqslant 1.5\,\text{m/s}$$

(this is a simplification since Δh is also affected by n).

(a) Determine the operating points for the pump for
    (i) a Newtonian liquid with $n = 1.0$
    (ii) a pseudoplastic liquid with $n = 0.9$
    (iii) a pseudoplastic liquid with $n = 0.8$.
(b) Comment on the effect of slight pseudoplasticity on centrifugal pump operation.

(4.4–1) A volute centrifugal pump has the following performance data at the best efficiency point:

volumetric flow rate Q	$= 0.015 \, \text{m}^3/\text{s}$
total head Δh	$= 65 \, \text{m}$
required net positive suction head $NPSH$	$= 16 \, \text{m}$
liquid power P_E	$= 14\,000 \, \text{W}$
impeller speed N	$= 58.4 \, \text{rev/s}$
impeller diameter D	$= 0.22 \, \text{m}$

Evaluate the performance of an homologous pump which operates at an impeller speed of 29.2 rev/s but which develops the same total head Δh and requires the same $NPSH$.

(4.4–2) Calculate the constant in equation (4.4–8) as a function of the specific speed N_s for a centrifugal pump given that

$$1 \, \text{U.S. gpm} = 6.3089 \times 10^{-5} \, \text{m}^3/\text{s}$$

and

$$1 \, \text{ft} \quad = 0.3048 \, \text{m}$$

(4.5–1) Two centrifugal pumps are connected in series in a given pumping system. Plot total head Δh against capacity Q pump and system curves and determine the operating points for
(a) only pump 1 running
(b) only pump 2 running
(c) both pumps running
on the basis of the following data:

operating data for pump 1

Δh_1, m,	50.0	49.5	48.5	48.0	46.5	44.0	42.0	39.5	36.0	32.5	28.5
Q m³/h,	0	25	50	75	100	125	150	175	200	225	250

operating data for pump 2

Δh_2 m,	40.0	39.5	39.0	38.0	37.0	36.0	34.0	32.0	30.5	28.0	25.5
Q m³/h,	0	25	50	75	100	125	150	175	200	225	250

data for system

Δh_s m,	35.0	37.0	40.0	43.5	46.5	50.5	54.5	59.5	66.0	72.5	80.0
Q m³/h,	0	25	50	75	100	125	150	175	200	225	250

(4.5–2) Two centrifugal pumps are connected in parallel in a given pumping system. Plot total head Δh against capacity Q pump and system curves for both pumps running on the basis of the following data:

operating data for pump 1

Δh m,	40.0	35.0	30.0	25.0
Q_1 m³/h,	169	209	239	265

operating data for pump 2

Δh m,	40.0	35.0	30.0	25.0
Q_2 m³/h,	0	136	203	267

data for system

Δh m,	20.0	25.0	30.0	35.0
Q_s m³/h,	0	244	372	470

(4.6–1) (a) Name some types of pumps which are seriously affected by mis-alignment.
  (b) What is the shape of the total head against capacity characteristic curve of a gear pump?
  (c) If very hot fluid is pumped with a gear pump, what difficulty might occur?
  (d) Gear pumps can be small liquid cavity high speed pumps or large liquid cavity low speed pumps. Which type would you use to pump
    (i) a pseudoplastic liquid?
    (ii) a dilatant liquid?
    (iii) a slurry?

(5.4–1) An agitator is used to mix a liquid of density ρ and dynamic viscosity μ in a cylindrical tank. The agitator has a diameter D_A and its rotational speed is N. The power P_A required to mix the liquid may be written as

$$P = f(N, D_A, \rho, \mu, g)$$

Use dimensional analysis to derive an expression of the form

$$\frac{P}{\rho N^3 D_A^5} = C\left(\frac{\rho N D_A^2}{\mu}\right)^x \left(\frac{N^2 D_A}{g}\right)^y$$

(5.5–1) Calculate the theoretical power in watts for a 0.25 m diameter, 6 blade flat blade turbine agitator rotating at $N = 4$ rev/s in a tank system with a power curve given in Figure (5.5–3). The liquid in the tank is pseudoplastic with an apparent dynamic viscosity dependent on the impeller speed N and given by the equation $\mu_a = 25(N)^{n-1}$ N s/m² where the power law index $n = \frac{1}{2}$ and the liquid density $\rho = 1000$ kg/m³.

(5.5–2) For laminar flow of a Newtonian liquid in a stirred tank, the power P_A is given by the equation

$$P_A = \mu C N^2 D_A^3$$

where

μ_a is the liquid dynamic viscosity

N is the agitator speed

D_A is the agitator diameter

and

C is a constant for the system.

A pseudoplastic liquid has an apparent dynamic viscosity given by the equation

$$\mu_a = K(N)^{n-1}$$

where the consistency coefficient $K = \mu$ at a power law index $n = 1$. Show that for the same power the pseudoplastic liquid can be agitated at a higher agitator speed N_1 given by the equation

$$N_1 = N^{2/(n+1)}$$

(5.7–1) Solute free liquid at a volumetric flow rate Q is used to purge off quality solute from a stirred tank of volume V. Show that if three equal size tanks are used in series, the removal of solute is n times more effective after a time t where n and t are related by the equation

$$t = \frac{V[(2n-1)^{1/2} - 1]}{Q}$$

(6.4–1) An ideal gas in which the pressure P is related to the volume V by the equation $PV = 75 \text{ m}^2/\text{s}^2$ flows in steady isothermal flow along a horizontal pipe of inside diameter $d_i = 0.02$ m. The pressure drops from 20 000 N/m² to 10 000 N/m² in a 5 m length. Calculate the mass flow rate in kg/(s m²) assuming that the basic friction factor $j_f = 4.5 \times 10^{-3}$.

(6.6–1) Calculate the linear air velocity in m/s required to cause a temperature drop of 1 K on a conventional thermometer given that for the air at atmospheric pressure and 373 K, the thermal capacity per unit mass at constant pressure $C_p = 1006$ J/(kg K).

(6.6–2) An ideal gas flows in steady state adiabatic flow along a horizontal pipe of inside diameter $d_i = 0.02$ m. The pressure and density at a point are $P = 20\,000 \text{ N/m}^2$ and $\rho = 200 \text{ kg/m}^3$ respectively. The density drops from 200 kg/m³ to 100 kg/m³ in a 5 m length. Calculate the mass flow rate in kg/(s m²) assuming that the basic friction factor $j_f = 4.5 \times 10^{-3}$ and the ratio of heat capacities at constant pressure and constant volume $\gamma = 1.40$.

(6.8–1) Show that the work required to adiabatically compress an ideal gas from a pressure P_1 to a pressure P_2 in a compressor with two equal stages is $[(P_2/P_1)^{(\gamma-1)/4\gamma} + 1]/2$ greater than in a compressor with four equal stages where γ is the ratio of heat capacities at constant pressure and constant volume.

(7.2–1) A mixture of gas and liquid flows through a pipe of internal diameter $d_i = 0.02$ m at a steady total flow rate of 0.2 kg/s. The pipe roughness $\varepsilon = 0.000045$ m. The dynamic viscosities of the gas and liquid are $\mu_G = 1.0 \times 10^{-5}$ and $\mu_L = 3.0 \times 10^{-3}$ N s/m^2 respectively. The densities of the gas and liquid are $\rho_G = 60$ kg/m^3 and $\rho_L = 1000$ kg/m^3 respectively. The weight fraction of gas is 0.149. Calculate the pressure gradient in the pipe using the Lockhart Martinelli correlation.

(8.2–1) A Pitot tube is used to measure point velocities in water. The reading on a mercury manometer attached to the Pitot tube is 1.6 cm. Calculate the linear water velocity given that the specific gravity S.G. $= 13.6$ for mercury.

(8.2–2) Calculate the volumetric flow rate of water in m^3/s through a pipe with an inside diameter of 0.2 m fitted with an orifice plate containing a concentric hole of diameter 0.1 m given the following data:
   (1) a difference in level of 0.5 m on a mercury manometer connected across the orifice plate
   (2) a mercury specific gravity S.G. $= 13.6$
   (3) a discharge coefficient $C_d = 0.60$.

(8.3–1) Calculate the volumetric flow rate in m^3/s through a V-notch weir when the height of liquid above the weir is 0.15 m given that the notch angle $\theta = 20°$ and the discharge coefficient $C_d = 0.62$.

(8.5–1) Show that a flowmeter with a square root scale has an error $50/Q$ times that for a linear scale where the maximum volumetric flow rate $Q_{max} = 100$ per cent.

(9.3–1) A gas of density $\rho = 1.25$ kg/m^3 and dynamic viscosity $\mu = 1.5 \times 10^{-5}$ kg/(s m) flows steadily through a bed of spherical particles 0.005 m in diameter. The bed has a height of 5.00 m and a voidage of $\frac{1}{3}$. The pressure drop is 150 N/m^2. Calculate the linear approach velocity in m/s.

(10.1–1) Calculate the time in seconds and in hours for a liquid to fall in a tank from a height $z_1 = 9$ m to a height $z_2 = 4$ m above a discharge hole of diameter $d_i = 0.02$ m given the following data:

   tank diameter $D_T = 2$ m
   dimensionless correction factor $\alpha = 1$
   gravitational acceleration $g = 9.81$ m/s^2
   the pressure over the liquid in the tank is equal to the pressure at the outlet.

(10.2–1) Nitrogen contained in a large reservoir at a pressure of 200 000 N/m^2 and a temperature of 300 K flows under adiabatic conditions through a horizontal converging nozzle with a 0.05 m diameter throat. The pressure at the nozzle throat is 100 000 N/m^2. Assume that the mean linear velocity in the nozzle throat is below the critical velocity. Also assume frictionless flow and ideal gas behaviour. Calculate the nitrogen flow rate in kg/s given the following data:

gas constant R_G = 8.3143 J/(kmol K)

molecular weight (MW) for nitrogen = 28.00 kg/kmol

the ratio of heat capacities at constant pressure and constant volume γ = 1.40.

(10.3–1) Show that the time to reach 50 per cent of the terminal velocity for a spherical particle falling from rest in laminar flow in a fluid is

$$t = 0.071u_p$$

where u_p is the terminal linear velocity.

(12.6–1) A Newtonian liquid of density ρ = 1200 kg/m^3 and dynamic viscosity μ = 0.001 kg/(s m) flows in steady state over a horizontal flat plate with a stream velocity of 3.0 m/s. Calculate the Reynolds number N_{RE} and the boundary layer thickness δ in metres at a distance 0.1 m along the plate from the initial point of impact of the liquid with the plate.

(14.6–1) A cylindrical vessel of inside diameter 1.0 m containing liquid is rotated to produce a maximum vortex depth of 0.05 m. Calculate the rotational speed of the vessel.

Answers to problems

(1.4–1) (a) $\mu_a = K(\dot{\gamma})^{n-1}$

(b) $\dfrac{\mathrm{d}\mu_a}{\mathrm{d}\dot{\gamma}} = -\dfrac{K}{(1-n)(\dot{\gamma})(\dot{\gamma})^{1-n}}$

(c) $\mu_a = 1.5\,\mathrm{N\,s/m^2}$

(d) $\dfrac{\mathrm{d}\mu_a}{\mathrm{d}\dot{\gamma}} = -3 \times 10^{-3}\,(\mathrm{N/m^2})\,\mathrm{s^2}$

(1.6–1) (a) $\dfrac{P_1 - P_2}{L} = \rho g \sin\theta$

(b) $L = \dfrac{P_1}{2\rho g \sin\theta}$

(c) $17\,\mathrm{m}$

(2.4–1) (a) $u = \left(\dfrac{\mu}{\rho d_i}\right)\left[\left(\dfrac{\Delta P}{L}\right)\left(\dfrac{\rho d_i^3}{0.1584\,\mu^2}\right)\right]^{\frac{1}{2}}$

(b) $1.10\,\mathrm{m/s}$

(2.4–2) $\mu = \left[\left(\dfrac{\Delta P}{L}\right)\left(\dfrac{\rho d_i^3}{64\,000}\right)\right]^{\frac{1}{2}}$

(2.4–3) (a) $\dfrac{\Delta P}{L} = \left(\dfrac{2\rho u^2}{d_i}\right)\left[4.06 \log\left(\dfrac{d_i}{\varepsilon}\right) + 2.16\right]^{-2}$

(b) $286.5\,(\mathrm{N/m^2})/\mathrm{m}$

(2.4–4) $\dfrac{\Delta P}{L} = Cf\left(\dfrac{\rho u d_i}{\mu}\right)\dfrac{\rho u^2}{d_i}$

(2.5–1) $1584\,\mathrm{N/m^3}$

(2.5–2) $293\,\mathrm{N/m^2}$

(2.7–1) $\left(\dfrac{-\mathrm{d}v_x}{\mathrm{d}r}\right)_w = \left(\dfrac{8u}{d_i}\right)$

263

(2.7–2)   confirm laminar flow
    (a) $v_{max} = 3.13$ m/s
        v_x (at $d_i/4$) $= 2.35$ m/s
    (b) 3×10^{-3} m³/s

(3.3–1) 1000 (N/m²)/m

(3.6–3) $N_{VIS} = \dfrac{(\mu_a)_{pw}}{(\mu_a)_{pm}} = \left(\dfrac{K_w}{K_b}\right)\left[\dfrac{2n}{(n+1)}\right]$

(4.2–1) 1.038 m

(4.3–1) $\Delta h = k_1 + k_2 Q^{1.75}$

    where the constant

$$k_1 = (z_d - z_s) + \frac{(P_d - P_s)}{\rho g}$$

    and the constant

$$k_2 = \frac{(\Sigma L_{es} + \Sigma L_{ed})(0.239)}{(g d_i^{4.75})}\left(\frac{\mu}{\rho}\right)^{0.25}$$

(4.3–2) (a) (i) $\Delta h = 6.88$ m, $u = 0.96$ m/s
          (ii) $\Delta h = 6.28$ m, $u = 1.21$ m/s
          (iii) $\Delta h = 5.60$ m, $u = 1.45$ m/s

(4.4–1) $D_2 = 0.44$ m
          $Q_2 = 0.060$ m³/s
          $P_{E2} = 56\,000$ W

(4.4–2) $3.23 \times 10^{-3} N_s$

(5.5–1) 250 W

(6.4–1) 621 kg/(s m²)

(6.6–1) 47.8 m/s

(6.6–2) 623 kg/(s m²)

(7.2–1) 2775 (N/m²)/m

(8.2–1) 1.99 m/s

(8.2–2) 5.42×10^{-2} m³/s

(8.3–1) 2.25×10^{-3} m³/s

(9.3–1) 2.315×10^{-2} m/s

(10.1–1) 4510 s
          1.253 h

(10.2–1) 36.6 kg/s

(12.6–1) $N_{RE} = 3.6 \times 10^5$
          $\delta = 2.9 \times 10^{-3}$ m

(14.6–1) 1.98 rad/s

Index

Velocity
   angular, 162, 247
   apparent mean linear in a packed
      bed, 163
   calculation of in a pipe, 23
   dimensionless, 34
   entrainment, 168
   falling, 159
   fluidisation, 167
   friction or shear stress, 33
   maximum point in a pipe, 28, 31
   mean linear, 13, 207, 221, 226
      in a pipe, 29
   minimum for slurries, 169
   point, 2
   settling, 159
   sonic, 112
   standard for slurries, 169
   terminal, 159
Velocity distribution
   correction factor, 13, 14, 30, 33,
      58, 107
   for laminar flow in a pipe, 27, 29,
      225
   for power law fluids, 54
   for turbulent flow in a pipe, 30, 31

Velocity distribution—*continued*
   universal for turbulent flow in a
      pipe, 33
Velocity gradient, 3
   in a pipe, 42
Velocity head, 13, 147
Velocity meters, 153
Vena contracta, 140, 142
Viscometers, 5, 44, 248, 249
Viscosity
   apparent, 5, 40, 41, 46
   apparent in mixing tanks, 95
   dynamic, 3
   kinematic, 4, 33, 182, 208
   Newton's law of, 2, 3
Voidage fraction, 161
Von Karman equation, 20, 45
Von Karman integral equation, 211
Vortexing, 86, 95, 249, 251

WEBER number, 91

YIELD number, 49
Yield stress, 5